The Magic & Romance of

LOS ANGELES, CALIFORNIA

This book is lovingly dedicated to
Our Parents
and to those who refuse to stop playing
and will never grow old.

GRAPHIC CREDITS

This book was photo-composed by Andresen Typographics, Tucson, Arizona.
The color separations were provided by Quadracolor, Burbank, California.
The book was printed by Dai Nippon, Tokyo, Japan.

Copyright © 1986 by Farago Publications, Inc. All rights reserved.

P.O. Box 48200, Los Angeles, CA 90048

Art Direction and Design: Stephanie Farago
Photography: Bob Dennison
Design Associate: Richelle Drake
Production: Lee Jordan
Production Associate: Dee Malan
Cover Design: Peter Palombi
East Coast Photography: Sandra Lee Kaplan
Printing and Color Consultant: Bert Roberts
Production Coordinator: Linda Girasole

From the Prologue by Carl Sandburg to the
Family of Man © 1955, ren. 1983,
The Museum of Modern Art, New York
All rights reserved reprinted by permission.

We have tried to be accurate and thorough in the presentation of this book. It may contain errors of commission or omission. We apologize for any difficulty they may cause our readers. We trust they will understand that the size of this project and human nature may have inevitably permitted some errors to elude us and are not substantial in nature.

All rights reserved under International and Pan-American Copyright Conventions.
Published in the United States by Farago Publications, Inc.

Library of Congress Catalog Card Number 85-080880
ISBN 0-935363-00-9

Manufactured in Tokyo, Japan
First Edition April 1986

The Magic & Romance of
Art Dolls

Text, Illustration and Design by
Stephanie Farago

Photography by
Bob Dennison

Produced and Published by
Fred Farago

ACKNOWLEDGEMENTS

The creation of *The Magic and Romance of Art Dolls* has been more than a labor of love, it has been our passion. During the six years it took us to assemble it, we were fortunate to meet many generous, loving and exceptionally brilliant individuals. Our meeting with Erté was an historic event which none of us will ever forget. His participation in this project made the use of the word "magic," in our title, essential.

Traveling in France, England, Monaco, Italy, Spain and the East and West Coasts of the United States, we were constantly delighted by the people and collections we were fortunate to see. This book would not be possible without any of the people we are about to mention ... finding someone to head this list is nearly impossible, since everyone included gave 100% to its birth and life. Our deepest gratitude to Richard Wright, Steven Arnold, Arnold Wells, Dorothy and Jane Coleman, John Noble, Gladyse and Arthur Hilsdorf, Ruth Noden, Madame Bordeau, Madame Channas, Sandra Lee Kaplan, Peter Palombi, Gregory Weir-Quiton, Marty Tunick, Jack Woody, Tom Long, Bob and Richelle Drake, Linda Girasole, Dee Malan, Bert Roberts, Lee and Sandra Jordan, Sophia Rodriquez, Ann Palmer, Harold Lipton, Laurence Glassman, Basil Collier, Jack, Adele and Ed Kennedy, Harvey Wine, Ed and Reina Lipnick, Raymonde Weintraub, Craig Butler, Lexy Scott, Terry Harstad, Jim Heimann, Bob Rodriguez, Roger Beerworth, Nicholas Farago, Don Pinnegar, Siri Singh Sahib Bhai Sahib Harbhajan Singh Khalsa Yogiji, Hari Har Kaur, Shakti Parwha Kaur, Soorya Kaur, Chakrapani, Sampuran Singh, Gretchen Schields, Robert Foothorap, Mieko Itchikawa, Barbara Wallasch, Herman and Anita Lipney, Dan Spelling, Paul Whitehouse, Tom Boland, Keith and Donna Kaonis, Regis G. Thomas, Stuart Claude Tilleard, Beppe Garella, Margaret and Blair Whitton, Barbara Whitton Jendrick, Bill Washburn, Frank Penn, Ruth West, Alain Labbe, Kerrick Foster, Jackie Petras, Shirley Buckholtz, Pat Gosh, Celina Carroll, Charlotte Dinger, Sheila Bradley, Ralph Griffith, Richard Saxman, Nancy Peterson, Arlene Adler, Jess and Bonnie Rand, Jeanette Fink, Ruth West, Liz Friedberg, Marge Bravin, Simone Harrington, Christian Gemignani, Ron Bakal, Louise Nemschoff, Clifford and Heather Bond, Countess Marie Tarnowska, Marsha Bentley Hale, Francois Theimer, Jacque Bouriquet, Ariane del Vecchio, Vespucci Puccini, Brian Doran, Francois Romanetti, David Solzberg, Loraine Burdick, Lucille McClure, Hector Perez, Tom Fronterhouse, J. C. Carr, Honeya Barth, Annette Annechild, Mary Wagner, Alberta and Marco Bossi, Richard Rheem, Tony Menchin, Deon Blue, Lenny, Sophie and Michael Haimovitz, Richard Linn, Susannah Farago, Marianna and Charlie Wallasch, Tamara Fago, Doug Rosenthal, Dolph Gotelli, Ellen Klugman, Anita Matthews, Romey Keys, Audrey Crandall, the Dennison family, Carla Folk, Jim Noyd, Nancy Carrasquillo, Robin Keats, Douglas and Jeri Duncan, Judith Kory and Darryl Paul, Rhonda Columb, Gert Leonard, Mickey Raskin, Estelle Johnston, Scott Brastow, Mel Birnkrant, Janie Norwood, Louisette Levy-Sousson, Kurt Triffet, Karen Pedersen, and in fond memory of Barbara Krieger.

Foreword

Dolls are rich and sensuous works of art which may be taken seriously without missing any of their intended delight.

The doll is humanity's oldest and most prevailing symbol of joy and amusement. Although dolls are the oldest sculptural archetypes, as a serious art form they have been largely ignored for the very fact that they are regarded as instruments of play. Our purpose in compiling this book is to reveal a wondrous and exciting part of our cultural heritage that might otherwise be missed.

Anthropological studies have shown every race, every culture, and every land had God, the family unit and dolls. As far back as 40,000 B.C., since the time of the Cro-Magnon man, throughout the entire breadth of Europe and Asia, dolls have accompanied the advancement of civilization.

Traditionally, art has been a serious subject. Up until the 1900's it was thought that art must be either beautiful or poignant in the very classical sense. Only recently has art been accepted and encouraged to be humorous, playful, disturbing, and inventive enough to allow for audience participation. However, doll art has always had the freedom to stimulate participation, express all emotions, incorporate sound, music, movement, and mechanical special effects.

It has also been traditional in the art world to have "pedigrees" and signatures to establish authenticity, value, and higher resale. Many of the dolls we have chosen to photograph do not have such traceable lineage. Their value comes from being just what they are: interesting, sometimes mesmerizing objects of art and love.

Some of the dolls pictured here are old, fraying, and disintegrating. One could not say they were anywhere near "mint" condition (perfectly maintained and original), yet some of the most ravaged by time make the most startling aesthetic impact. Internationally known antique and doll dealer, Richard Wright, says such dolls have "Teddy Bear quality," that over the years they have assumed a patina, a look of having been intimately held and loved by their owners in such a way that truly touches the soul.

Dolls are romantic symbols. Rudolph Valentino is the picture of masculinity, Columbine has an aloof coquettishness, while the Cubeb Smoker is a sultry siren. They engender a world of sentimentality as they re-create déjà vu again and again. Once you fall in love with a doll's face, every time you see it, it's like seeing an old friend. Entering a room with over a thousand dolls, their languid long limbs gracefully draped over sofas and chairs, and meeting with 2000 poetic eyes is a unique and captivating experience. Each expression whispers "I love you," "Love me," asking "Do you understand me?" "Let me entertain you," "Please enjoy me," creating a feeling of intimacy and tenderness captured in time for our lasting pleasure.

Until now, the topic of "art dolls" from the Art Deco 1920's and 1930's has been vir-

Foreword

tually untouched. Mysteries surrounding dolls of this era will inevitably unfold as interest continues to develop. Perhaps more information will surface as heirs to collections sort through inheritances, memorabilia, and antiquated piles of paper.

Art dolls, also known as boudoir dolls, were a phenomenon of the Art Deco Era. Since this kind of long-legged doll has experienced somewhat of a rebirth in recent years, it is our pleasure to share some of the old ones that we have been fortunate enough to acquire or have had permission to photograph. We have not, in all cases, been able to find examples of each and every kind of doll manufactured by companies mentioned. If it is possible, in our lifetime, perhaps we will.

This book is a collage of dolls, times, eras, fashions, and fantasies. In creating the elaborate backgrounds and sets for the dolls, we drew from our own sources, friends, and prop houses. It was fortunate we never had a garage sale. Just when we would think we did not have an important element to complete a set, it would somehow always manifest itself. In some cases we have mixed and matched in a creative rather than an historical way. Dolls are complimented by backdrops similar to those used in movie sets, including original matte paintings and illustrative/photographic superimpositions. If anachronisms exist, they are there because they enhance the styling.

Our intention has been to create a visually dynamic book which would always be entertaining and informative. We also hoped to impart, in some small way, the undeniable grace and enchantment that dolls continue to bring to people's lives.

Introduction

Erté, at home, standing before one of his "Formes Picturales," sculptural reliefs which he created with various materials: duraluminium, copper, iron and wood, circa 1960.

Introduction

OUR MEETING WITH ERTÉ

On January 30, 1984, we had the fortunate opportunity to meet, just outside of Paris, with one of the most distinguished artists and stylists of this century, known throughout the world as Erté. Among his vast contributions to fashion and design over the past eight decades, Erté is recognized as the foremost innovator of the Art Deco style. Particularly significant to the fashion industry, he is considered to be the first fashion designer to illustrate his own creations.

In Erté's studio, accented by exotic animal furs, shells, fish, purring cats, cooing doves, his images of hand-painted metal, editions of bronze sculptures, hundreds of original designs and graphics, we spent an enlightening afternoon as we encountered the vitality of Erté's still great mind. At the age of almost 92, he lucidly recounted social mores, fashions, fads, attitudes and experiences that continue to shape our world.

—SF

Remembering back to the time of the salon dolls (boudoir dolls) is not an effortless task, since the past is something I don't give much thought to. I am always looking to the present and future to see how I can be of some value in making things better.

Of course, I do remember that I invented doll handbags as early as 1918 and 1921, along with doll muffs and doll-headed hat stands, and incorporated dolls as symbols into one of my color paintings in February 1919, when I was illustrating for *Harper's Bazaar*. I have always felt that dolls, if they are beautiful, should be displayed as works of art.

The dolls you see in these pages are some of the best examples of the surviving cloth lady and men dolls and are truly works of art. They capture the essence of "joie de vivre" inspired by the freer dance and clothing generated by the theatrical and operatic influences of the early 1900's. They highlight and dramatize the romance, mystique, glamour, imagination and fantasy of that time.

I feel that, historically, the preservation of the salon doll and all dolls is important because they exhibit fashion through the ages. The one thing I have always disliked in fashion is when it becomes like a uniform, inhibiting individuality. A wonderful feature of these dolls is that they actually burst free of previous molds. During the 1920's, romance and freedom reigned together while prudery and hypocrisy took a back seat to beauty, sensuality and grace. Men designed for women to liberate them with colour and freedom of movement, while concurrently freeing themselves to wear more colorful and interesting ties, patterns and fabrics.

I have always been dedicated to life, liberty and the pursuit of happiness and currently I have an even greater involvement, since I am participating in the restoration of one of the most charismatic women in sculpture...The Statue of Liberty. I have created my own version of her to sell with some profits going toward her eventual repair. It is my hope that this project, this book and others like it will keep the enchantment and elegance of the past alive.

Erté designed covers for Harper's Bazaar *1915 — 1936. Early work shows use of dolls as poetic symbols (right). The woman represents the United States playing an imperialistic game with dolls representing international countries.*

"Les Poupées Russes," by Erté, February 1919 cover Harper's Bazaar/Hearst Corporation. Courtesy © Sevenarts Ltd., London.

Eager eye and willing ear,
Lovingly shall nestle near.
In a Wonderland they lie,
Dreaming as the days go by,
Dreaming as the summers die:

Ever drifting down the stream
Lingering in the golden gleam—
Life, what is it but a dream?
—LEWIS CARROLL, ALICE IN WONDERLAND

Table of Contents

A 1920's Lenci Harlequin foretells the future. Farago Collection.

Table of Contents

FOREWORD

INTRODUCTION BY ERTÉ

I   A BEDTIME STORY FOR ADULTS

The meaning and symbolism of dolls through the ages.

II   IT'S NO LONGER CHILD'S PLAY WHEN A DOLL SITS ON AN ADULT'S BED

The evolution of adult dolls from French fashion mannequins to Art Dolls from France, Germany, Italy, England and the United States circa 1920 – 1935.

Movie star memorabilia, French post cards, and dramatic influences on the dolls of the Flapper era.

III   LENCI, VIRTUOSO OF THE BOUDOIR DOLL

Biographical/pictorial documentary on one of the most incredible doll empires ever known.

Table of Contents

IV   BRING IN THE CLOWNS

       Harlequins, Pierrots, and Clowns.

V   OTHER ADULT AMUSEMENTS

       Accessories, automata, other oddities.

VI   ENCORE

       The fine art of dollmaking—a continuing tradition.

       INDEX

"The doll is the highest and most complete form of human playlife... More complete than art, more complete than science is man's expression of himself through the symbolism of his own image.
—DR. GEORGE DAWSON, 1919,
    "THE PSYCHOLOGY OF DOLLS"

I. A Bedtime Story for Adults

The Pierrot Writer, an automaton by Vichy, 1875. Courtesy Musée National, Collection de Galéa, Monaco.

Art Dolls

Visions attributed to the fertile minds of children are actually fantasies designed by grownups. Engraving by Jules Taverner.

Pretending is something grownups do best. Originally, dolls and fairytales were never intended for children. *Grimm's Fairy Tales* were actually grim. The first dolls, dating back to the Ice Age, were not childlike in any way. Like the first folk tales, they were primal expressions of basic human emotions, possessing serious and dramatic social significance. Though dolls have now become an intrinsic part of the child's world, it is the intention of this book to focus instead on dolls created of, by, and for adults.

Throughout history, adult dolls featuring mature characteristics and proportions have had many uses, from mystical to mundane. Dolls reflect values, customs, trends, fashions, and notable personalities. They have been instrumental in religion, politics, courtship and countless other areas of life. Although adult dolls have always existed, their presence over the past 150 years has been partially eclipsed by the proliferation of baby-faced dolls first introduced during the Industrial Revolution.

It was not until the 1900's that the adult doll came into its own as a commercially popular art form. These bonafide adult art dolls appeared in the United States as early as the 1890's and were later adopted and developed to a more sophisticated level in England, France, and Germany. They were known alternately as "art dolls," "salon dolls," "boudoir dolls," and "bed dolls." Boudoir dolls was the most popular term, with cloth construction being their most distinguishing characteristic. By exploring the romantic overtones of the boudoir and its "offspring," the boudoir doll, we will reveal how this intriguing art form developed.

Due to the crippling economic effects of the Depression, the heyday for this type of doll (1900 – 1930) was short lived. However, some of the most elegant and elaborate sculptural forms in doll history were produced during that time.

For the last century, the production and sale of dolls has never enjoyed more popularity than it does today. Doll collectors comprise the second largest group of hobbyists in the world. Over 20,000 members belong to the continuously growing American organization, the United Federation of Doll Clubs.

Recently, there has been a resurgence of popularity of adult dolls inspired by the movie stars of the twentieth century. These

doll portraits include Greta Garbo, Marlene Dietrich, Joan Crawford, Humphrey Bogart, Mae West, Charlie Chaplin, Marilyn Monroe, and other multifarious sex symbols and famous characters.

Greta Garbo, Twenties Goddess. Her beauty inspired trends in fashion and dollmaking from the Art Deco Era to the present.

Madonna (right): German icon, circa 1800, with human hair. Nearly every major charismatic character, whether biblical, historical, or rock and roll, is depicted in the doll world. Farago Collection.

fashioned dolls in the image of woman to symbolize the forces and circumstances which affected them.

These early dolls were mainly fertility symbols or religious artifacts. Max Von Boehn, in his 1929 book *Dolls and Puppets*, attributes the preponderance of the female in doll sculpture to "the communal standard of centuries based on the 'mother-right'; the mother [occupied] the higher role in the social order; the conception of the father's importance [had] not yet developed."

This renaissance of dolls patterned after adults can be appreciated more fully by exploring how the first dolls came into being. Few dolls created prior to 1815 still remain intact, because materials used in earlier dollmaking could not withstand the test of time. Painting and literature are excellent sources in many cases, however, one must leave sheer conjecture to fill obvious gaps when exploring the inception of the first dolls. By examining the likely motives of prehistoric societies for inventing dolls, it becomes more evident why dolls have continued to be an important part of humanity's creative expression.

Most likely, primitive dollmaking was not a form of narcissism with women sitting around the cave fashioning likenesses of themselves. After all, those were not the days of women corporate executives, leisure time and evening university extension classes for women developing alternatives to housewifery. Those were the days of sustaining life against enormous odds. Men were the hunters; women the nesters.

Venus of Willendorf, a paleolithic limestone fertility idol, originating circa 40,000 B.C. Credited as the first doll image. Curiously, the female form predominated throughout doll history.

DOLL ORIGINS...MYSTIQUES: BIRTH, DEATH, WOMAN, MAN

Mystery immediately confronts us when researching the history and symbolism of dolls. The most intriguing fact is that most of the early dolls were fashioned in the image of women. Voluptuous feminine forms were developed as early as 40,000 B.C. Artists

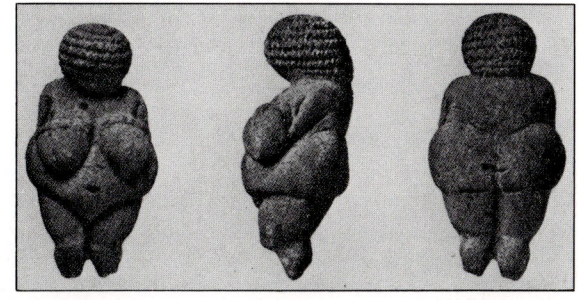

The woman seemingly held all the magical power of procreation and nurturing with womb and breasts man would never have. Early man probably viewed women as possessing supernatural power. In the role of observer, he produced sculptural forms, documenting the miracle of procreation and immortalizing his woman, the symbol of fertility and source of new life.

WOMAN AS IDOL/GODDESS

It is apparent how woman came to be worshipped as a goddess. A woman's fertility and endurance were her dowry in a world where the safety of medicine and hospitalization did not exist. After every birth, the unpredictable perils of disease, exposure, preying animals, or invading tribes took their toll on the birth rate. Woman became much like the goose that laid the golden egg: not only highly desirable, but indispensable.

DOLL POWER—Early dolls served as the objects of many magical and heathen rituals such as adulation, healing, voodoo, fertility, satanic rites, witchcraft, protection against the evil eye, and guarding against perdition or the loss of one's soul.

Voo-doo dolls, ancestral figures used for revenge, exhibit one of the innumerable functions of dolls.

A sculptural rendition of Botticelli's Venus, the quintessential goddess of love and art. Woman's beauty has been celebrated historically as goddess/idol/doll.

Realizing the serious intent of these early dolls, it is clear that these were not toys for children; certainly no concerned prehistoric parent would allow his child to play with them. Dolls fulfilled a major function by helping to comfort man and insulate him against fear. While one of man's greatest fears has always been the inevitability of death, he created images in his own likeness in an attempt to immortalize himself and leave evidence of his existence. By creating images and gods in his own likeness, with infinite, supernatural power, he became equally empowered.

The first "gods" had human qualities and represented the gods of fire, rain, love, and others. Some of these anthropomorphic gods became known as Zeus, Venus, Jupiter, Mars, etc. By ascribing human characteristics to them, man was able to visualize them as all powerful men and women. He could emulate them, participate with them and appeal to them to help solve his earthly dilemmas. Greek mythology illustrates man's use of sculpture to endow

Courtesy of The Detroit Institute of Arts. Founders Society Purchase, Mr. And Mrs. Benjamin Long Fund. Miscellaneous Gifts Fund and City of Detroit Insurance. Recovery Fund.

Pygmalion myth exemplifies man's desire to be God-like in creating and bringing to life his perfected images.

"*How good—to be alive! How infinite—to be, Alive—two fold—The Birth I had— And this—besides in—Thee!* —EMILY DICKINSON

himself with god-like attributes. In the myth *Pygmalion*, a lonely king of Cypress carved an ivory figure of a beautiful maiden. He became so enamored of his handiwork that he prayed to Venus, Goddess of Love, to transform the statue into a real woman. When his prayers were answered, it verified his ability to compel the gods and, inevitably, be god-like himself.

This impulse to immortalize through the use of model figures is ancient. The earliest graves have revealed dolls serving as companions or servants to the deceased in afterlife. In Tutankamun's tomb, a male bust, used as a tailor's dummy, was among the objects found when the grave was opened in 1922. Roman custom provided for funerary figures in the likeness of the corpse; Julius Caesar's funeral is said to have featured a wax statue, complete with twenty-three bleeding stab

A tailor's dummy, representing King Tut (above) was discovered in his tomb in 1922. Regarded as the first portrait mannequin.

wounds. In the Middle Ages, European aristocracy elevated this practice to an art, the earliest example of which is the wax-headed replica of King Edward III of England, who died in 1377.

From these funerary images we can trace the evolution of mannequins both as portraits and as functional items used by tailors and artists. Michelangelo and Albrecht Dürer both used lay figures as models. Life-sized wax mannequins weighing up to 300 pounds were used as a medium for portraiture by the European nobility. Says mannequin authority Marsha Bentley Hale, "Each of these forms of human imagery had its own specific influence upon contemporary display mannequins."

DOLLS AS DIPLOMATIC EMMISSARIES AND FASHION COMMUNICATORS

Surprisingly, dolls have played an important role in communication at all levels of society, from the royal courts to the bourgeoisie. European nobility frequently used dolls as gestures of friendship and to promote fashion trends. The chain of events surrounding England's King Richard II and his ties with France indicate that using dolls had political impact.

In 1391, during peace negotiations between France and England, Queen Isabella of France may have actually changed the course of history when she sent a life-sized doll, a mannequin, presumably as a gift of goodwill to the English Queen, Anne of Bohemia. This figure is regarded as history's first female display mannequin. Four years later, when Queen Anne died, her husband chose to marry a French princess. Perhaps inspired by Isabella's clever approach, King Richard II used another doll as a means of wooing his bride-to-be.

In 1396, King Richard II commissioned his court tailor, Robert de Varennes, to design a doll's wardrobe, presumably a custom-fitted trousseau, for his betrothed. It is speculated that the doll used to transport the garments to England may have been another life-sized French-made mannequin because of its exhorbitant price. This historic episode indicates that the diplomatic gesture of sending a doll may have succeeded in cementing relations between the two countries.

Since there were no fashion periodicals until 1670, when *Mercure Gallant* was published, dolls served as communicators to advertise and sell new designs. The development of fashion dolls flourished in Europe; Paris and other cities became centers of dollmaking. Small, jointed mannequins called "Little Pandoras" were used to show casual fashion, while larger models, "Big Pandoras," displayed formal wear. The French Pandora dolls were also known as *grand couriers de la mode*. However, the name Pandora was chosen for its symbol of irresistibility, since these high fashion models were used to charm customers into purchasing the clothing as well as the dolls themselves.

Pandora myth (left): Zeus punished Prometheus for his theft of fire by sending the irresistible Pandora to earth with a box full of the ills of mankind. The Tate Gallery, London.

French wax fashion doll (right), circa 1900, represents fashionable forerunners of the boudoir dolls of 1910 — 1935. She wears 17th century dress made from authentic antique fabric, 23" high. Musée National de Monaco, Collection de Galéa.

Pandoras were named after a woman characterized in a Greek myth in which Zeus punished Prometheus for his theft of fire. He ordered Hephaistos to create a woman with all the attributes necessary to enchant men. The temptress Pandora was sent to earth with a box containing all the ills of mankind, which escaped when her curiosity compelled her to open the box. The name Pandora was chosen for these high fashion modeling dolls to encourage sales with their symbolically seductive qualities.

Traces of Pandoras have been found for centuries across the Western world. In the 1400's, Queen Anne of Brittany commissioned a large Pandora doll to be sent to Queen Isabella of Spain, showing her the current French fashion. In 1600, Henri IV, the French king known for his ruggedness and equivalent odor, attempted to gain the affection of his intended, Marie de Medici, by sending her a number of small French fashion dolls. It is interesting to note that during the Napoleonic blockade of 1803, small wax Pandoras were allowed passage over borders when people and parcels were refused. Dolls have been used universally during wartime to smuggle secrets and contraband.

EVOLUTION OF THE FASHION DOLL

It is difficult to ascertain exact information about many of the doll forms that existed, since so few prior to 1815 have survived. At that time, dolls were crafted individually by hand. Some dollmakers were employed as journeymen while others made dolls as a sideline, or for their own enjoyment. Since there were no patents, dating the dolls accurately can present a problem. Therefore, paintings and literature have become the main reference materials. Dating costumes is also a way to determine when dolls were made, but often the costumes would disintegrate with time and be replaced. In his book *Histoire des Jouets*, author Henri d'Allemagne covers the history of dolls from the 1300's to the 1600's. There are no pictures, though the dolls are described in the text.

While dolls for adult purposes were the most prevalent for hundreds of years, it cannot be denied that children naturally would create their own and that some dolls made by adults were intended for children. Since the distinction between child and adult is not always clear, it is not known for whom many of the dolls were made. In most cases, however, well-crafted dolls were owned only by children of royalty or the very rich. For the common man, whose main concern was survival, such luxuries were superfluous.

Children prior to the 1700's rarely had time for play; they were dressed and treated exactly like adults. For them, life was a serious business. Often marrying young and dying early, they were expected to behave and work on a mature level. For the most part, dolls for children were used as educational tools, preparing them for adulthood; unclad dolls gave little girls an opportunity to learn how to sew clothes, dress fashionably, and hopefully make them desireable for marriage. Furthermore, the use of dolls as children's playthings was undoubtedly discouraged in many instances because of evidence of poisonous paint that was used on a great number of dolls that existed prior to 1800.

Fashion doll (left), circa 1880, Jumeau typifies the kind of dolls mothers gave their daughters to encourage familiarity with fashion and etiquette, and to better prepare them for marriage. Liz Freedberg Collection.

Maury Album, circa 1815, illustrates how perishable dolls during that era may have looked.

Lady fashion dolls became more prevalent during the time of Marie Antoinette and her dress designer, Rose Bertin. Throughout the 1700's until the French Revolution, dolls were made of wood or wax, with painted or glass eyes. A great variety of dolls were made at this time. By the middle of the 19th century, mannequins or tailors' dummies were used and probably displayed in the store windows of the Rue St. Honoré, where Rose opened her shop. She and the poets of that period praised the French mannequins known as the "dolls of the Rue St. Honoré."

The *Maury Album*, a doll catalogue dated 1815, contains wonderful illustrations of the Empire Period (1804 — 1815). Many of the faces drawn resemble Napoleon Bonaparte's son. The drawings pictured are quite beautiful, but in actuality the dolls salvaged from that period are quite crude. With rounded, wooden faces and painted or glass eyes, they are almost two-dimensional, except for the noses.

Napoleon's son (right), born in 1811, purported to be the subject of doll illustrations in the Maury Album.

Grödnertal dolls and pegwoodens made in Germany began appearing in the middle of the 1700's. Because they are wooden, many have endured the test of time. Queen Victoria enjoyed collecting these dolls.

Molded papier-mâché dolls from Sonnenberg-Thuringia in Germany were the first to be mass-produced around 1850. French dolls known as *carton*, a kind of papier-mâché or cardboard, may have preceded the examples made from the Regency (1811 – 1820) and Empire (1804 – 1815) Periods. Dates are difficult to pinpoint, however, since few of the carton (molded paper) dolls have survived.

Grödnertal (left) is a wooden doll, 1815 – 1840, and the carton (right) a type of paper construction. Both are forerunners of the long-limbed lady dolls of the Deco Era. Musée National de Monaco, Collection de Galéa.

LOUIS XIV, MASTER OF ARTIFICE AND THE POLITICS OF FASHION

Fashionable dolls created during the days of Louis XIV's elegant French courts belonged to a class of luxury items meant only for monarchs and their wealthy friends. They were gifts given with the same intention as the Fabergé eggs—as prizes, rewards, and tools of seduction. These baubles of brilliance and craftsmanship were pawns in the game of the politics of fashion. What must be realized is that if something happened in fashion, it also inevitably happened in the doll world.

Louis XIV, the Sun King 1643–1715, established Paris as the fashion capitol of the world. All dolled-up, he grew from 5' to nearly 6' with wig and high heels.

Elegant dolls, like Fabergé Eggs, were given to gain political advantage and social favors.

From <u>The Art of Carl Fabergé</u> by Kenneth Snowman.

Louis proclaimed France the fashion headquarters of the world, and it remained so until the first royal head rolled for indulgence in regal excess. He coined the phrase "Fashion is the mirror of history," and created a domain where his nobles were expected to dress in elaborate opulence at their own great expense. By demanding this attention to fashionable extravagance, Louis kept the courtiers enslaved to his every desire. They were forced to constantly ask him for favors, since they were always in debt from their unending expenditures.

Just over five feet tall, he grew to nearly six feet after donning pointed high-heels and a grandiose powdered wig. All dolled-up, he was the master of artifice and always surrounded by it. Continuously in the company of his elegantly glittered leeches and their ladies, he was not even permitted to sleep alone. Whenever he and the Queen had intercourse, one of the attendants was required to wear his hat a certain way to signify the

consumption. Records were kept in order to determine if and when he actually impregnated her.

Versailles was glorious and boring. Dressing and undressing for balls, masquerades, and orgiastic banquets was the court's major occupation. There are court dolls—male, female, and hermaphroditic. (We know this because they are fully sexed. Certainly these dolls must have spiced up court life). The dolls attributed to this period have been challenged as reproductions because their arm joints seem to be too recent to have come from Louis' court. However, these dolls are truly representative of the ones created during Louis' time.

Court dolls, two views: (left) hermaphroditic, (right) male, anatomically correct. All wood, 10" high, possibly from Louis XVI's court. Clifford and Heather Bond Collection.

The courtiers had no privacy, always having to be "on." This, coupled with the great competition to out-do each other in their daily get-ups, resulted in great affectations, hypocrisy, tension, scheming, and snobbery.

In Italy, royalty dealt quite well with the stress of constant adulation. They designated a special place in the palace garden, called the *piajare*, where no one could hear them. There, they could retreat to solitude and scream as loud and as long as they liked.

FROM SALON TO BOUDOIR: THE DERIVATION OF THE BOUDOIR DOLL

In the 14th century at the Chateau de St. Germaine, the princesses had what they called a cabinet, or small room, just outside the bedroom. Eventually, a *salon* (room) just before the entrance to the bedroom appeared in the architecture of a number of regal environments. Called the *boudoir*, taken from the root "bouder," to pout, it became the place where one could go and finally just be oneself.

The Marquise de Rambouillet in 1618, Ninon de Lenclos, and other women of the court had another civilized strategy for coping with the claustrophobic, male-dominated court life. The Marquise gathered her contemporaries together in the first celebrated salon which lasted for over half a century. Having established a need for a private place to entertain her most intimate friends, the Marquise created her salon

Ninon de Lenclos helped in establishing the first Parisian salons. The boudoir and the salon were interchangeable meeting places where women and men could socialize romantically and intellectually. Musée Royaux des Beaux-Arts de Belgique, Bruxelles.

Ben Franklin was a frequent visitor at Parisian salons.

on the Rue Saint-Thomas-du-Louvre. On certain days, selected members of the bourgeoisie, artists and intellectuals—including Voltaire, Molière, Balzac and Benjamin Franklin—would be invited. Descartes wrote his *Treatise on Love* there.

Upon entering the salon, one might be coaxed into joining the conversation with such provocative greetings as "Pray, what do you think of love, madame?" or "In your opinion, what does the soul consist of?" Somewhat of a precedential feminist movement, their elite clique demanded a woman's right to comport her life as she chose, to select her own mate and to freely pursue the goals of cultural and spiritual satisfaction.

MADAME DE POMPADOUR
"... The first great interior decorator."

Madame de Pompadour was the subject for many fine portraits by painters Boucher and Watteau. The memory of her grace and charm was also immortalized in doll form by the famous Lenci Company in the early 1920's (see back cover photo). Not only was her beauty great, but so was her eye for talent. The artists and craftsmen that she patronized during her presence at Versailles gave tribute to her through the artifacts they created. She contributed an elegant, creative brilliance which affected the works of artisans for generations. Without her influence on the art of decorative motifs, it is possible that dolls and artifacts might never have evolved to their standard of elegance.

Madame de Pompadour was another fashion trendsetter. Probably the greatest interior decorator of all time, she was the force behind the acceptance of Orientalia and the subsequent Rococo style. She could dance, sing, act brilliantly, and play the clavichord; she was a gardener and botanist. Moreover, she was a fine artist and could paint, draw and make engravings. Never known to lie, she was noted for her incredible charm and her frank sense of humor.

Madame de Pompadour's corset is on exhibition at the Musée de la Femme in Neuilly, France. The corset is so tiny, that it is nearly

impossible to imagine her fitting all those attributes into such a diminutive frame! It has been commented that Madame de Pompadour might possibly have died from her practice of partying until 3:00 a.m. each

Madame de Pompadour, known as the first great interior designer, was the subject for many fine paintings as well as the Lenci doll. Painting by Boucher, National Galleries of Scotland. Pompadour (below) all felt Lenci doll, height 25", circa 1927. Farago Collection.

morning and having to appear at Mass refreshed and beautifully attired again only five hours later.

The trend of the salon spread throughout France. An exciting retreat, quite innovative for those times, the styling of the Marquise de Rambouillet's salon became a part of French architecture. It was an honor to be received by the Marquise as she held court from her grand bed in the glorious "Blue Room." Combined with creative brilliance and elaborate styling, her salon became known as the first recognized "boudoir" in architecture.

The boudoir was more than just a room. With centuries of development, the consequent naming of the "Boudoir Doll" held more significance than just a doll sitting on a bed. It was the richness, romance, and creativity of the environment of the boudoir that truly inspired the name for this special doll.

The boudoir became an inner-sanctum, where the lady of the court could take off her restricting corset, wear her "saque" dress, a loose, flowing gown, and commiserate with her king, converse with intellectuals and statesmen, or just throw her tantrum in peace. Elegantly appointed with all her favorite gifts, dolls and possessions, she could mix her magic potions and practice her rituals of beauty without anyone knowing of the effort behind her fresh disguise.

The only respite from tedious hours spent at Versailles in exceedingly uncomfortable boned corsets and trains, came in the blessed form of "at home" or "undress" clothes—flowing gowns accessorized with short, fur-trimmed jackets and frivolous little capes.

Before becoming known as Madame de Pompadour, her name was Antoinette Poisson, or Miss Fish. Straight from the bourgeoisie, she consulted a fortune-teller, at the age of nine, who told her she would "reign over the heart of the King…" In gratitude, Madame de Pompadour left this woman 600 livres in her will.

Surprisingly, it was her expertise in driving that caught the eye of Louis XVI. One day, while the handsome king was on a hunt, she strategically rode her horse-drawn blue phaeton across his path. She was wearing a pink gown that she had made herself. The very next day, she reversed her ensemble, driving a pink phaeton and wearing a blue gown. Her distinctly feminine flair captured his heart and the hearts of many.

Madame de Pompadour contributed a sense of style in everything she did. In her boudoir at Bellevue, although politics were not one of her strong points, she may actually have had an influence over policies made within her charming domain. She may indeed have inspired the term "boudoir diplomacy," which either means that great international decisions were made within the walls of the boudoir or that a man could sweet-talk his way into the arms of a lady in the boudoir. If he were a fast worker, he could have her on the way to the bedroom. Certainly, the latter was not the case with Madame de Pompadour, since she was madly in love with Louis.

Madame was frail and was often forced to withdraw to the comfort of her boudoir, this specially designed suite, originally created for pouting and brooding, may very well have been a sanctuary for her in her frequent bouts with illness and feminine dis-

Commedia dell' Arte characters (front left to right) Pierrot, Columbine and Harlequin exhibit costumes which originated during the Renaissance and continue to inspire dollmakers to this day.
—Porchoir print by Barbier.

comfort. It is said that the palace residents were so entrenched in etiquette that they may have even worn diapers, being too self-conscious to excuse themselves to use a toilet. No water closets were found at the Palace of Versailles...only privies in the courtyard. The boudoir may have been the only private retreat available to nobility.

Loved and respected by the most powerful man in France, accepted almost unanimously by both court and commoners, Madame de Pompadour was the most successful courtesan in history and an example for women everywhere. She entered the palace in 1745 equipped only with her talents and charm. Having no vocation, wealth, or prominence, she soon rose to a position of premier influence in the royal court.

Marriage, until the 20th century, was one of the only acceptable vocations for a woman, except for dressmaking. Before the arrival of the couturier, every woman, including Madam de Pompadour, was required to create her own fashions. Unable to marry the man in her prophecy, since Louis was already married to the Queen, Madame did the best thing. Far from being a consolation prize, she became "the woman behind the throne." She received all the benefits of being a mistress without the domestic fallout.

CELEBRITY DOLLS:
CHARISMATIC CREATIONS

The art of handling fabric and becoming accomplished as seamstresses or designers were the only avenues of recognition available to women. These sublimated souls began to excel in the world of creation, adornment, and amusement. In history, there were numerous talented women painters, but they were never included in art history books. Clearly, their path was to be cut out of the cloth afforded to them. Dollmaking was a perfect playground for talented European women, especially since women made up the vast majority of doll buyers.

Family companies in Germany and England were creators of porcelain pincushion dolls and accessories. These sensuous and seductive figurines were most likely the creations of the male members of the family who were admirers of the female form. The tactile pleasure of sculpting these lovely nudes and elegant ladies must have brought these men creative solace in a time when men were expected to look, but not touch. These companies were Dressel-Kister, Kestner, Goebels, Royal Doulton, Royal Dux and others.

The theatre, stemming from 15th century Italian Commedia dell'Arte to ballet, opera, and eventually the cinema, offered fertile territory to creative craftsmen and women. Ladies of the night took their cues from women like Madame de Pompadour,

1920's factory women cut their careers out of the fabric afforded to them.

Yvette Guilbert (left) by Toulouse-Lautrec. A trendsetter in her time, her long black gloves were her trademark. Musée Toulouse-Lautrec, Albi.

Ninon de Lenclos, and the Marquise de Rambouillet. These women realized themselves as the coquettes, grande cocottes and the demimonde, the kept women of the upper classes. As social barriers diminished, these mistresses became independent, wealthy and proud. In the Montmartre, women like Yvette Guilbert, La Goulue, and others became the central figures for Toulouse Lautrec's impressionistic paintings. From chanteuse to actress, women were contributing to the visual splendor of art nouveau. Around the turn of the century, the theater was about the only place a woman could openly express herself. She might be disowned by her family, but not by her public.

In Paris, Sarah Bernhardt was immortalized by Mucha, the great poster artist, and by doll artists Mlle. Lifraud and Claude Marlef. Lifraud made the wax head and Marlef painted it. The Sarah Bernhardt doll is

© 1923 by Familia Media, Inc. Reprinted with permission of Ladies' Home Journal.

Sarah Bernhardt doll (left), as Camille, is French, silk-faced, height 34"

Sarah Bernhardt poster (right) is one of many designed by Alphonse Mucha, the famous poster artist. Bernhardt inspired Mucha as well as Parisian dollmakers.

part of the Union des Arts collection which includes portraits of these other celebrated Parisian actresses: Mme. Rejane, Segond-Weber, Bartel, Sorel, and Granier.

Meanwhile, England's school of Pre-Raphaelite painters blazed through Victoriana with their crimson-haired models, igniting a torch-like glimmer in a world burning to be born.

The Bridesmaid by Mallais, Pre-Raphaelite beauty inspired dollmakers.
The Fitzwilliam Museum, Cambridge.

King Edward VII, Queen Victoria's son, didn't set much of an example when he took the red-haired "Jersey Lily," Lily Langtry, as his mistress. Lily was sculpted in wax as a fashion doll, possibly made by Lucy Peck.

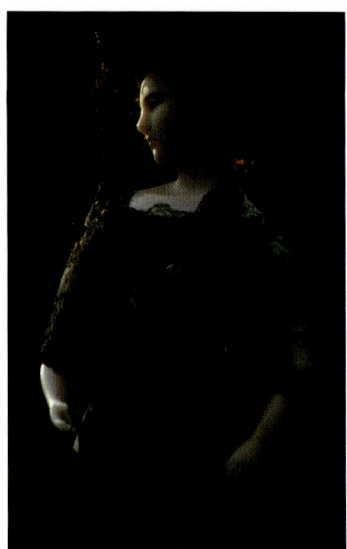

"Lily Langtry" doll (right), perhaps by Lucy Peck, circa 1900, height 22", wax head and arms with Pre-Raphaelite red hair. Clifford and Heather Bond Collection.

Women were coming out of "the closet"... out with everything they had held back for years. The philosopher Nietzsche remarked that as most real men liked two things: danger and play—he would choose women, "the most dangerous plaything." Oscar Wilde, whose philosophy "nothing succeeds like excess," was also noticed out of his closet when he was imprisoned for homosexuality... "love that dare not speak its name."

Oscar Wilde (above), Irish dramatist, libertine philosopher-poet and wit. On his death bed he looked at the wallpaper and remarked, "Either the wallpaper goes, or I do."

Oscar Wilde doll, presumably English, height 26", papier-mâché head with cloth body. Gladyse Hilsdorf Collection, Fayetteville, N.Y.

Cocottes like La Paiva in Paris and the cockney, Cora Pearl, commanded as much as ten thousand francs a night from their lovers. With society no longer intimidated by the tyranny of "etiquette" pamphlets and "what people might think," corsets and bustles both went out like lights.

Art Dolls

VENUS TO VAMP

Squeezed by prudery long enough, dealing the death blow to the industry that made 50 million corsets a year, Paul Poiret felt he was merely the first to respond to women's inner longings to be physically, therefore spiritually, liberated. He saw himself as satisfying these obvious desires and not imposing his own willful trends. He made almost everyone happy when he replaced the old contraption with the newer models...the brassiere and the rubber girdle.[1]

Leon Bakst's costuming (below), for "Scheherazade" and other Ballet Russe productions, influenced Art Deco styling.

Isadora Duncan's freer dances demanded freer wraps. She influenced designers to release women from restrictive dress. "Isadora Duncan—Her Life, Her Art, Her Legacy" by Walter Terry.

As early as 1628, Harvey's discovery of blood circulation aroused suspicion about headaches, fainting spells and other pains resulting from the wearing of corsets. By the Gay 90's, some women had their ribs surgically removed to accomplish an 18" waist, while others actually died from tight lacing. American sociologist, Veblen, claimed the corset was an immobilizing garment, causing mutilation and glorified servitude, since it prevented any kind of constructive movement, like work.

Typical corset as seen in 1900's advertising.

The dances of Loie Fuller and Isadora Duncan, in the 1900's, expressed a physical and emotional freedom never seen before. All Paris was infatuated with Diaghilev's Ballet Russe production of *Schéhérazade*, starring the incomparable Nijinsky, with sets and costumes by Leon Bakst. This extravaganza began a trend in which anything exotic became the rage. Poiret and his wife Denise, dressed in turbans and Arabian pantaloons, gave a "1002nd Nights" costume party. The Poirets entertained their guests in an Ali Baba Cave, complete with peacocks, flamingos, parrots, monkeys and lanterns of fruit lighting up the night.

Nijinsky's grace (above) greatly influenced Deco designers.

Poiret designed high-fashion clothes for Lafitte-Deserat dolls. He made exotic themes fashionable and liberated women from corsets.

---

[1] Paul Poiret's claim to fame is rivaled, however, by Germany's Otto Tizlinger who is said to have designed the brassiere to replace the corset of the famous opera star, Swanhilda Oloffson, for a particular virtuoso performance in which she appeared.

La Fitte-Deserat dolls, 1912 — 13, dressed by Poiret, were given away when customers made expensive purchases. Wax heads and arms, with cloth bodies, human hair wigs, 12" high. The Strong Museum, Rochester, N.Y.

In 1917, Poiret was designing doll clothing for Lafitte-Deserat wax dolls. Soon after, he advocated women carrying dolls as fashionable accessories, dressed in identical costumes to match exactly what the ladies were wearing.

Lanternier (left), 1915 — 16, French bisque swivel head, glass eyes, cloth body, height 15". Another forerunner of the Deco lady dolls. The Strong Museum, Rochester, N.Y.

Reclining, all wax doll, 11" long, is less of a fashion indicator and more of a doll. The Strong Museum, Rochester, N.Y.

A healthy, colorful glow was applied to the faces of fashion, theatre, opera, ballet and the arts décoratifs, soon to be known to the world as Art Deco. Erté, who had been influenced since childhood by Persian paintings and dramatic theatrical events, illustrated women with exaggerated Deco doe-eyed mystery, which eventually led to a total revolution in make-up for women. Eventually, Erté became a designer for the great Parisian mannequin maker, Pierre Imans, while concurrently contributing to the phantasmagoria of the French theatre. For the 1913 performance of "Le Minaret," he designed a dress for one of the dancers from the Ballet Russe, the infamous Mata Hari.

Erté's costume design for Le Minaret, 1913, for Mata Hari. © Sevenarts Ltd., London.

In 1914, the Industrial Revolution caught us speeding up

Pierre Imans' French wax mannequins, made during the 1920's, were unmistakably works of art. They were fabulously stylish and amazingly life-like although they weighed as much as 200 lbs. Erté sculpted mannequins for Imans, see page 210. Photograph: Marsha Bentley Hale Collection.

to the door of World War I. Ladies just set free for frivolity were also free to serve as paid employees in the war effort. Somehow, the sickly spectre of war only encouraged the festivities. Freud, the "father of psychology" and author of *The Interpretation of Dreams*, also fathered the use of cocaine, while Cocteau was experimenting with opium.

Romance and escapism offered a hopeful relief to the nightmare of war. Facing destruction, people turned to fast cars, the silver screen, and one last tango, which seemed more important than ever before. During the Armistice, couples in small Parisian restaurants began the practice of dancing between courses.

The Roaring Twenties set the pace: automated, fast and free. Advertising, under the influence of Hollywood and yellow journalism, proclaimed miracle cures and make-up promising fountains of youth and beauty to equal the stars. With child-like rouged cheeks, short skirts, and long-legged lady dolls poised upon their naked knees, the flapper women reached for the beauty of innocence in a bizarre and decadent way.

From basinette to boudoir, the flapper era brought about the most enigmatic of all dolls. While its doll form suggested child's play, the personality of the creation was all sophistication and vampishness. Adorned with high-heeled shoes, exotic-lidded eyes, and a cigarette dangling from a ruby red mouth... "The French doll... triumphant in the fashionable dress and millinery shops and apartments, is long, startingly swank, and elegant, but surely not for children." From the salon doll, the boudoir doll and the bed doll, an art doll emerged as the mature plaything in a grown-up aficionado's life.

To become mature, one must achieve that sense of seriousness of a child at play."

—FREDERICK NIETZSCHE

II. It's No Longer Child's Play When a Doll Sits on an Adult's Bed

The Harlequin Honeymooners, American Blossom dolls (circa 1920's) in an amorous mood, celebrate the heyday of Art Deco dolls for adults of all ages.

Hollywood's greatest Art Deco doll is still the Oscar, photo by Bob Dennison. Still photo of set, Henry Alvarez Collection, Courtesy Twentieth Century Fox.

Art Dolls

> *Joan Crawford is doubtless the best example of the flapper, the girl you see at smart nightclubs gowned to the apex of sophistication... dancing deliciously, laughing a great deal with wide, hurt eyes.*
> —F. Scott Fitzgerald

From the MGM release "Our Dancing Daughters"
© 1928 Metro-Goldwyn-Mayer Distributing Corporation.
Ren. 1956 Loew's Inc.

Joan Crawford and her look-a-like doll that she brought to be used in the film "Our Dancing Daughters" in the late 1920's.
From the MGM release "OUR DANCING DAUGHTERS" © 1928 Metro-Goldenwyn-Mayer Distributing Corporation, Ren. 1956 Loew's Incorporated.

The magic of movies revealed beautiful people, beautiful manners, and beautiful fashions. Theda Bara, the first "Vamp", Mary Pickford, "America's Sweetheart", Pola Negri, "The Magnificent Wild Cat", Clara Bow, the "It" Girl and others, provided women all over the world with the stereotypes they could spend their lives aspiring to be. Fashions swept off the screen and onto the streets almost overnight.

People were fast becoming aware of what was "in" and what wasn't. "IT" was "in." "IT" was "attitude", an elusive quality you either had or you didn't. "IT" was glamour, sex appeal, sophistication and free spirit. Thanks to "IT," sex was no longer considered unmentionable in a polite society. Women had long wanted to be sexy and now, for the first time, they had permission to flirt, to look interested and even to be provocative.

Art Dolls

The "IT" girl—Clara Bow

"IT" WAS IT, AND EVERYBODY WANTED "IT"
Of course, there was an "IT" Doll.

"'IT'—A strange magnetism that attracts both sexes, a virile quality, full of confidence, indifferent to the effect he or she is producing, uninfluenced by others. There must be physical attraction, but beauty is unnecessary. Conceit or self-consciousness destroys 'It' immediately."
—Elinor Glyn

In 1926, Elinor Glyn, an English writer coined the term "IT". She named Clara Bow the "IT" girl and ironically didn't think Gloria Swanson had "IT". She consequently and quickly wrote a novelette and script for a cool $50,000 called "IT" to further capitalize on her sensational promotional scheme.

Clara Bow was probably the Marilyn Monroe of her day. The titles of her films indicate just how much liberation was tolerable in 1924—*Daughters of Pleasure, The Adventurous Sex, My Lady's Lips, Parisian Love,* and *Kiss Me Again.* The girls with "IT," the stars and their fans, were known as flappers, vamps, and coquettes. Like Clara Bow, they "flaunted their disdain for conventional dress and behavior." These women were the first to shed their tight corsets for loose, comfortable pastel-tinted lingerie. Relieved of their victorian modesty, many of these women were now dressing up just to go to bed.

Clara Bow exemplified the consummate flapper. Coming from a tragic family life and total poverty, Clara's success story appealed to every ordinary American. Common people everywhere were scheming to get rich quick, be a Horatio Alger or be a "star." Suddenly media, publicity, and promotion could make one famous overnight...like fairy tales coming true.

Lady dolls created during these early film days were influenced by the cinematic flapper era. These long-legged lady dolls, cigarettes dripping from their lips and spit curls glued to their cheeks, were the "bad girls" of the doll world. Wonderfully provocative and a little bit weird, they were interesting and had character but were not really the type of

Anita Dolls (above), composition-headed bed dolls, lay in their boxes, tied in knots, smoking neurotically while they waited for a handy woman to dress them. They were the most prevalent bed dolls produced in America. Farago Collection.

Art Dolls

Probably one of the last corset ads shows Deco dolls with their lounging ladies who appear to be loosening up. Illustration by Ruth Wilcox.
Courtesy Vanity Fair © 1923 (renewed) 1951, 1979 by the Condé Nast Publications Inc.

girls you would want your children to play with or your sons to marry. Bad girls with hearts of gold, they were completely irresistible and loved by all. The vamp image was so popular that Betty Boop starred as the cartoon vamp of the silver screen. She was patterned after the original "Boop-Boop-A-Doop Girl," singer Helen Kane.

Betty Boop was the cartoon vamp of the silver screen. This Betty Boop silk/satin boudoir doll, 29" high, was reportedly made for Paramount Pictures. Farago Collection.

Dolls became so influential during the 1920's that the film "Cytherea" featured a doll named "Cytherea" as the focal point for the entire film.

39

Art Dolls

"The Blue Angel" (1929), starring Marlene Dietrich, featured Lenci dolls. Her beauty inspired Lenci dollmakers.

When silent stars went travelling abroad, they returned embracing armfuls of elegant, long-legged French silk-faced and Italian felt dolls. These exquisite treasures became the toys of the very rich. Soon American companies formed to imitate and capitalize on the fad of these romantic foreign dolls being bought by and for adults.

These dolls found abroad, during the early 1900's, were called boudoir dolls or bed dolls. They were the direct descendants of the French wax mannequins, fashion dolls dating back to the 1600's (when they were used to display new fashions prior to fashion illustration). In France, Lafitte-Deserat dolls (as seen in Chapter I), about 12" high, were costumed by Paul Poiret in 1910, while others (approximately 30" high) were dressed by Paquin and Mme. Lanvin in the early 1900's. Often these dolls were given away as a gift when a customer made an extravagant purchase.

The lady dolls created during the early 1900's sold for a variety of prices, depending on their origins and the materials they were made from. The early European dolls ranged from $40 to $200. In 1907, *Playthings Magazine* mentioned a new doll which was a "society woman of great elegance." In 1910, the *Ladies Home Journal* described a Paris doll for $150. In 1911, *Playthings Magazine* wrote about the Paris modistes (designers) who costumed lady dolls. The article pictured four of the "Dolls of High Degrees" and priced them at $40 to $50 apiece.

Renée Adorée with her look-alike doll on Photoplay *magazine cover illustrated by Carl Van Buskirk.*

Claire Windsor (below), an outstanding doll enthusiast, claimed to own over 200 boudoir dolls.

From the reference files of Lorraine Burdick. Celebrity Doll Journal, May 1967.

* © Universal Pictures, a Division of Universal City Studios, Inc.
Courtesy of MCA Publishing Rights, A Division of MCA Inc.

Art Dolls

Portrait dolls of the favored stars were beginning to appear. The fans were buying up these collectible souvenirs to savor while the stars were collecting dolls mimicking themselves. Dolls were sculpted after the Gish sisters, Rudolph Valentino, Gloria Swanson, Marlene Dietrich, Josephine Baker, Joan Crawford, Greta Garbo, Charlie Chaplin and others. Celebrities were fond of posing with their dolls for publicity.

Among the movie stars who collected these glamorous dolls were Marlene Dietrich, Joan Crawford, Claire Windsor, Dolores Del Rio, Lupe Velez, Colleen Moore, Carmel Myers, Gloria Swanson, Renee Adoree, the Gish sisters, Shirley Temple and Jane Withers. Claire Windsor was known to have collected over 200 boudoir dolls. Many of the more exotic dolls found their way into the homes of celebrities such as Ernst Lubitsch, the famous German director.

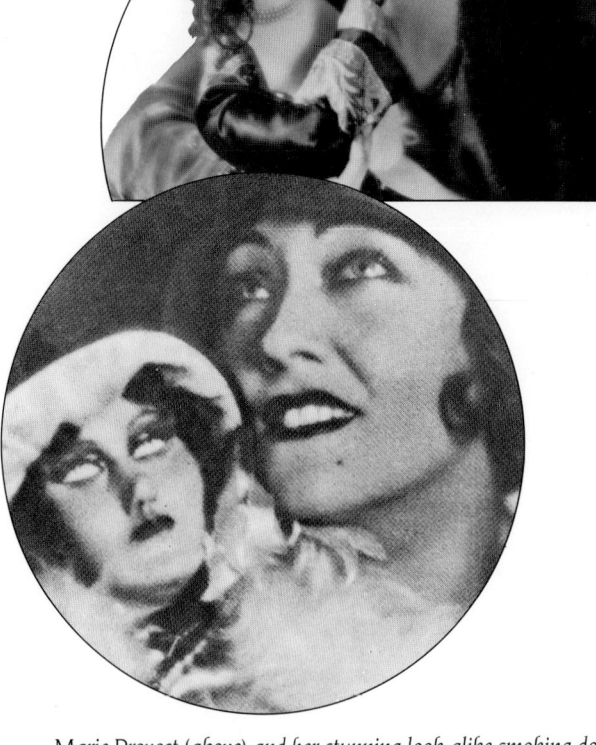

Marie Provost (above) and her stunning look-alike smoking doll.

Gloria Swanson (below) exhibits her radiance and her portrait doll. Reproduced with the permission of the Estate of Gloria Swanson. From the reference files of Lorraine Burdick. Celebrity Doll Journal, May, 1967.

Celebrities were fond of posing with their dolls.

Often the dolls would be used in film backgrounds for ambiance. In the film *Cytherea*, a doll named Cytherea was actually the focal point on which the story was based. Joan Crawford brought her own doll into the film *Our Dancing Daughters*. In Marlene Dietrich's first film, *The Blue Angel*, her co-star Emil Jannings, plays with a black Lenci doll. There are photos of her with a black doll and a Lenci Oriental boy doll Ms. Dietrich is said to own.

Lupe Velez also owned a significant doll collection.

From the reference files of Lorraine Burdick. Celebrity Doll Journal, May 1967. Courtesy of the Academy of Motion Picture Arts and Sciences.

Art Dolls

Nude photographs were politely termed "art photos." Here a coquette poses with her art doll.

Art Dolls

Among all dolls…the boudoir doll occupies a most special place, the bed. The doll has the most special of abilities— the power of transporting its owner to a time of personal life long past.

Many times the boudoir doll replaced Fido as the owner's only confidant.

Twenties women were openly flirting and revealing themselves behind a loose veil of silk lingerie and thin chemises from bordello to high fashion.—Richard Rheem

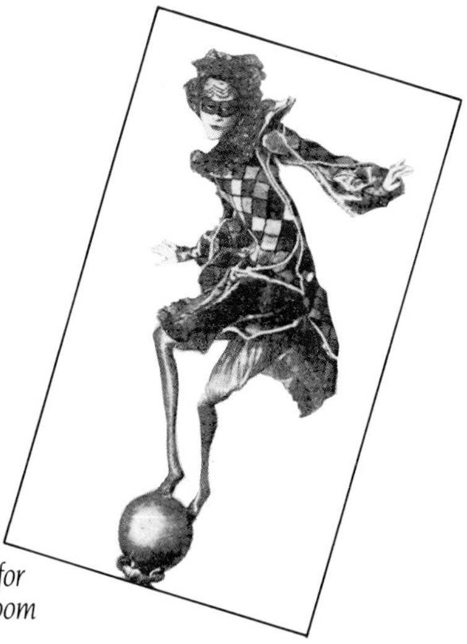

German Lotte Pritzel was famous for her dramatically posed "drawing room dolls." Truly innovative, her wax-headed dolls with cloth covered wire armatures could not be rivaled. Max Von Boehn says of the drawing room doll "...this type of doll first became art under the hands of Lotte Pritzel ...her dolls have more of the essence of our age than a whole glass palace full of modern pictures."

Max Von Boehn wrote about artists and doll-makers seriously developing art dolls in craft circles in the early 1900's for artistic competition and museum exhibitions. In 1921 he referred to an exhibition of art dolls held in Nürnberg, describing Lottie Pritzel's work as "[coming] to life like improvisations of the subconscious... [her figures embodied] all the peverseness of a soulless, hopeless species drowning in sensuousness."

Madame Lazarsky was Polish; she set up a stable of artists and other Polish immigrant workers to make these charming lady dolls.

Danseuses by Consuelo Fould —flexible lady dolls created in the early 1900's. Courtesy of Musée Roybel-Fould de Courbevoie, Paris.

THE RAVCAS

It is rare when two artistic individuals of the opposite sex marry, work together and support one another's creativity. This is the case with the wonderful couple known in the doll world as Bernard and Frances Deicks Ravca, who were both dollmakers before they met. They have since continued to work independently and collaboratively.

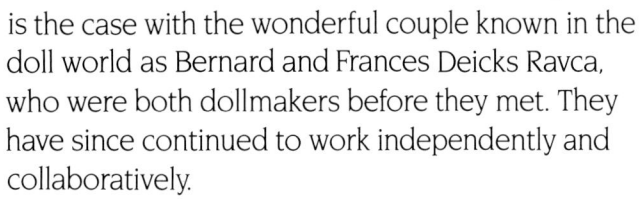

Both NIADA artists, Bernard's dolls are soft sculptures of cotton, covered with nylon stockings and hand-painted; Frances is known for her cotton needle structure.

Neville Chamberlain and Maurice Chevalier by Bernard Ravca.

The famous Lydia Pinkham doll by Frances Deicks Ravca.

Shop window displaying a variety of boudoir dolls and other cloth dolls, circa 1930. Playthings Magazine.

Most of these names came from a translated article entitled "La Renaissance de la Poupée Française," by Jean Doin, printed in the Gazette des Beaux Arts, 1916. Doin noted the popularity of yellow and black dolls (probably referring to Oriental and Negro). The article listed early 1900's European doll makers.

DECO/EUROPEAN DOLL MAKERS
- Kasparek
- Mme. Frankowska
- M. Sicard
- Mme. Ciechanowska
- Mme. Berthe Noufflard
- Mme. Deffands
- Mme. Roig
- Mme. d'Eichthal
- Mlle. Verita
- Mlle. Koenig
- Mme. Henriquez
- Mme. de Felice
- Mme. Yvonne Gall
- MM. Masson
- Mme. de la Boulaye
- M. Gardet
- M. Lejeune
- M. Marque
- M. Gumory
- M. Antoin
- M. Antoinin Mercie
- Union des Arts
- Mme. Zambelli
- Mme. Bartet
- Mme. Maillard
- Pallez
- Amaury
- Vernhes
- Guillaume
- Mme. Lifraud
- Mlle. Rozmann—faces looked very real
- M. Botta—Papier maché
- M. Georges LePape—Marionettes
- Mme. Lauth-Sand, granddaughter of George Sand
- Danerval de Laffranchy Co.—1914
- Marguerite Steif—German
- Mme. Lazarski—Polish, 1914
- Mme. Fiszerowna—Polish
- Mme. Ambroise Thomas—Known to have improved craftsmanship of the cloth doll at that time, especially the hair, hands and feet.
- Edward J. Siefert—German. Specialized in Harlequin manufacturing. From 1891 – 1920, he produced Marottes (doll heads on sticks) of Harlequin and Columbine.
- Mme. Alexandrowicz—Known for astonished faces, in wood.
- Mme. Desaubliaux—Her dolls were described as very French.
- Lafitte-Deserat 1910—Fashions Paul Poiret. (Mme. La Baronne de Laumont probably funded Lafitte-Deserat in 1914.)
- Mme. Manson was inspired by Dessault's tales.
- Mme. d'Aulnoy and Mme. Vera Ouvre—Their dolls were thought to be more modern than their contemporaries. Ouvre was credited with making a doll of "Charlot"—Charlie Chaplin.
- M. Bricon—Italian theatre inspired his figure of "Polichinelle"—Punch.
- Mme. Sweicka—Dolls known for their tenderness.
- Mme. Dhomont—Specialized in clowns and fantasy.
- Mme. Duvall—All her dolls smiled.
- Mme. Lloyd—Specialized in naive faces. (Duvall and Lloyd worked together and developed detailed and elegant underwear.)
- Mme. Claude Marlef—Did dolls of Bernhardt and Carlotta.
- Zambelli—Theatre and opera stars.

EUROPEAN DOLL COMPANIES
- Dean's Rag—England
- Chad Valley—England
- Norah Wellings—England
- Farnell—England
- Harwin—England
- Lenci—Italy
- Messina/Vat—Italy
- Alma—Italy
- Steiff—Germany

EUROPEAN DOLL ANECDOTES
British boudoir dolls were called names like Natty Nora, Captivating Cora, Tantalizing Flora and Florrie the Flapper.

Dean's produced cloth dolls until 1938; Chad Valley called them Carnival Dolls in the 1930's. Later in 1935, Chad Valley produced 21″ high "Sofa dolls... playthings of the Sophisticated Miss".

These Deco period dolls, the same as all dolls throughout history, reflected the fashion and trends of the time. They featured rosebud lips and side-glancing eyes with dramatic make-up, which sometimes included lush lashes. Their extremely long limbs duplicated the chic look of elongation found in fashion illustration. Elongation in styling produced an elegant quality not found in reality. Unlike any previous dolls, they had many more functions from decorating to advertising.

Dolls dressed as Commedia characters with Deco ladies, 1920's etching by Nandy.

McCalls offered clothing patterns for boudoir dolls. Pattern numbers 1776 and 1828 © McCall Pattern Company.

Considered more like art objects than dolls, they were sold in the decorator part of the store. They were called "beautifiers," displayed on the bed, sofa, table, or limousine and they were even carried along as an accessory for the evening. In 1911, *Toys and Novelties Magazine* reported "fine dolls have become the companions of milady—in the boudoir, in the restaurant, and at the theater, as well as in the automobile." Some eccentric German women were known to take rubber dolls into the baths with them.

Another phenomenon found in the doll world is the pincushion doll or half doll (with head, torso, and arms in one piece) which was also part of the boudoir doll era. Besides the fabric lady dolls used to enhance home decor, a variety of accessories using porcelain and composition art dolls were produced prior to World War I through the 1930's.

The pincushion dolls were modeled after elegant ladies, Pierrot and Pierette, dancers and other graceful themes. The imaginations of the boudoir doll makers went on without limit as they incorporated doll heads and half dolls made of porcelain, composition, and fabric into lingerie bags, party favors, hat stands, purses, radio dolls, calendars, candy boxes, whiskbrooms, telephone covers, lamps and other items as seen in Chapter V.

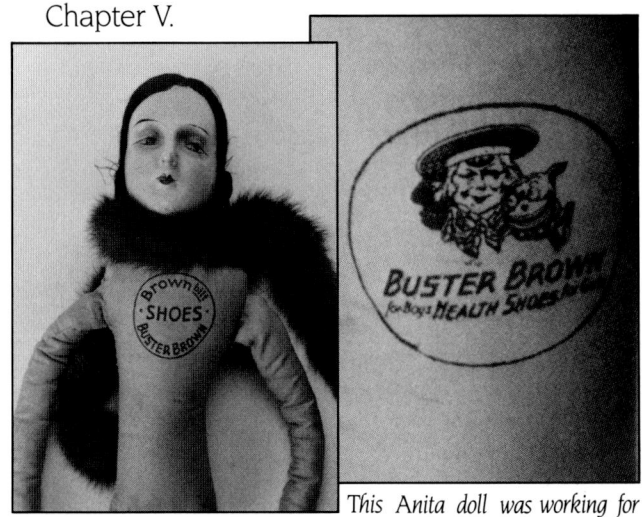

This Anita doll was working for Buster Brown Shoes, displaying printed logos front and back.

A 1920's illustration of a lady at play at her boudoir dressing table.

Nature is in earnest when she makes a woman.
—O.W. Holmes, *The Autocrat of the Breakfast Table.*

Claire Windsor speaking into her novel accessory, a radio doll phone.
From the reference files of Lorraine Burdick. <u>Celebrity Doll Journal</u>, May 1967.

American Art Dolls were also called Boudoir or Bed Dolls, Salon Dolls, Flappers, Vamps, Bed Dolls, Sofa Dolls, Wobblies, Costume Dolls, Bye-bye Kitties by Horseman, Phonograph Dolls 1878 — 1925, Cigarette Dolls, Parisians and Hoopla Girls, and Whoopee Dolls. Some came fully attired in gowns laden with black jet, baroque gilded laces, multi-layered frilly organza, or metallic diamond-shaped patterns silk screened right on their long arms and legs.

Younger women, who were anxious to grow up and show off their sophistication by being a part of every new fad, bought the exotic lady dolls. Older women were equally motivated to keep the blush of youth and were attracted to the dolls as symbols of youthful frivolity. Frederick Nietzsche said "To become mature, one must achieve that sense of seriousness of a child at play." Perhaps this explains the flappers' fascination with these remarkable artifacts. They were exploring the magnitude of their sovereignty as "romantically liberated" women in the 20th century. The devastating effects of the Crash of 1929 marked an early end to this era of creative fantasy. Lonely dolls sat forsaken on shelves, as doll manufacturers closed their doors when luxury items could no longer be afforded.

AMERICAN DOLL COMPANIES

Blossom
Etta
Cubeb
Gre-Poir
Konroe Merchants
Gerling Toy Company
Sterling
Flapper Novelty Doll Co.
American Stuffed Novelty Co.
European Doll Manufacturing Co.
Regal Doll Manufacturing Co.
Charles Bloom and Fred K. Braitling
Adler Favor and Novelty Co.
Mutual Novelties
Louis Amberg
Alfred Munzer, Inc.
Konroe Merchants
New York Sales Co.
Anita Novelty Co.
Unique Novelty Doll Co.

Flapper showgirl sits on her boudoir table.

AMERICAN DOLL COMPANY ANECDOTES

Gerling Toy Company, which marketed the "Whoopee" doll, also offered a set of flapper George and Martha Washington dolls for $2.98 each.

Fred K. Braitling sold boudoir doll shoes as well as dolls.

The Blossom Doll Company was financed with $5,000. Blossom dolls were all fabric, hand-painted faces, and came with and without eyelashes. It is suspected that they may have sent to Europe for the fine faces and attached them to their American-made bodies.

Etta was an all-women doll company that created fabric dolls with long eye lashes; owned and named for Mrs. Etta Kidd.

Cubeb, all composition and fully jointed, was a doll designed by a Russian immigrant, Samuel Haskell. The doll patent was applied for on April 25, 1924 and was one of the first "smoker dolls" created to encourage women to smoke Cubeb cigarettes, a menthol smoke touted to be an herbal remedy. Mutual Novelty and Konroe Merchants, both listed at the same address, offered this doll in their *Playthings* ads. It is possible these companies were one and the same.

The Cubeb Smoker and the Fadette, by Lenci, resemble each other so greatly, that it is certain that one inspired the other. The Lenci Fadette was documented in the 1922 issue of *Vanity Fair* while the Cubeb's patent was applied for in April 1924.

The "It" doll was created by Louis Amberg and Son and was called a Twist doll. It was all composition and was jointed at the waist so that it actually twisted. The skirt had a Twist label, and the pocketbook said "It". The "It" doll was accompanied by a puppy sitting with a leash on.

Regal Doll Manufacturing Co. introduced their "Lindy Doll" of Charles Lindbergh in February 1929.

Anita Novelty Company's bed dolls are the most prevalent bed dolls available today. A German wax doll may have been the prototype they imitated. The Anita Novelty Company collaborated with the European Doll Manufacturing Company of New York to produce special pillows with French heads designed by Altbuch. Together they created flapper doll heads to adorn boxes, lingerie bags and other accessories.

The European Doll Manufacturing Company of New York displayed fourteen different heads and five flapper dolls in their salesroom.

...soaps, skinfoods, lotions, hair-preservers and hair-removers, powders, paints, pastes, pills that dissolve your fat from inside, bath salts that dissolved it from without, instruments for rubbing your fat away, foods that are guaranteed not to make you fat at all, machines that give you electric shocks, engines that massage and exercise your muscles... A face can cost as much in upkeep as a Rolls Royce.

—CAROLYN HALL, THE TWENTIES IN VOGUE

© Conde Nast Publications Limited 1983

Ruby Keeler, known to some as the best tap dancer in the world, "tripped the light fantastic" with her Cubeb Smoker, 1928.

The Cubeb was one of the first "smoker dolls" created to encourage women to smoke Cubeb cigarettes, a menthol smoke touted to be an herbal remedy. Haughty and almost defiant, she reflects the chic 1920's attitude.

For years there has been discussion as to which came first, the Cubeb or the Lenci Fadette, since some are dressed in almost identical pant suits. The patent for the Cubeb was applied for in 1924, and a 1922 <u>Vanity Fair</u> features a photo of a Fadette (see page 122).

Indeed, there were no laws in either Britain or America to prevent copywriters from making fanciful or even blatantly outrageous claims for their products—Lucky Strike cigarettes, for instance claimed to protect the voice and eliminate coughing.

—THE TWENTIES IN VOGUE
© Conde Nast Publications Limited, 1983

GOTTA LIGHT? A *Cubeb Smoker* catches the eye of a gentleman passing by. Cubebs were designed by Samuel Haskell, a Russian immigrant. She is all composition with fully jointed elbows, hips and knees which enable her limber body to be posed in a sensuous and saucy manner, 25" high. Her face is hand-painted, and her hair is made of silk threads. Farago Collection.

The background painting is by Stephanie Farago and is a portrait of photographer Bob Dennison. It was, in fact, their first collaborative effort.

> The New Woman of the Twenties not only behaved and thought differently, she looked like a new species.
>
> —CAROLYN HALL, <u>THE TWENTIES IN VOGUE</u>
> © Conde Nast Publications Limited 1983

THE CASTING COUCH. *This doll resembles a number of silent screen stars, including Russian Nazimova and Theda Bara, the original "Vamp." She is probably French, has a hand-painted cloth face and cloth body, celluloid hands and sewn-in mohair wig, 30" high. Farago Collection.*

> Yet with low words she greeted me,
>   With smiles divinely tender:
> Upon her cheek the red rose dawned,—
>   The white rose meant surrender.
>
> —JOHN HAY, "The White Flag"

NEW YEAR'S EVE. She is an all cloth, silk-faced, French boudoir doll and bears a striking resemblance to a film personality...some say Dietrich. She has a hand-painted cloth face with a black velvet patch on her cheek (patches were used as early as the Renaissance as a decoration or to cover a pimple), all cloth body, and she is 30" high. Farago Collection.

Valentino postcards show the dollmakers' attention to detail
when they reproduced the costume.

... the face of Valentino was causing suicides; that of
Garbo still partakes of the same rule of Courtly Love,
where the flesh gives rise to mystical feelings of perdition.

—ROLAND BARTHES

THE SHEIK OF ARABY, Rudolph Valentino, whose real name was Rudolfo Alfonzo Raffaelo Pierre Filibert Guglielmi Di Valentina d'Antonguolla. Valentino is an Italian cloth boudoir doll with hand-painted chamois face, all felt body, 29" high, with label on bottom of foot crediting "La Rosa Company, Milano-Corso Venezia." Farago Collection.

...when two people are at one in their inmost hearts,
They shatter even the strength of iron or of bronze.
And when two people understand each other in their
   inmost hearts,
Their words are sweet and strong, like the fragrance of
   orchids.

—THE I CHING

WEDDING PARTY. Made by the Blossom Doll Company, one of the only American companies whose dolls are labelled. They were part of a 1930's wedding ceremony and may have been made especially for the extravagant event. After the wedding, they were packed away until 1985 when they were sold in absolutely mint condition.

They are all cloth, hand-painted silk faces, with eyelashes and mohair wigs. Wedding party includes bride, groom, best man and two bridesmaids, 30" high, two girl attendants, 24" high, and ring-bearer and Shirley Temple flower girl, 18" high. Farago Collection.

...one novelty followed another. These included pogo sticks, crossword puzzles, yo-yos, mah-jongg, bridge and potato crisps—one million packets were sold when they were first introduced in 1928.

—CAROLYN HALL, <u>THE TWENTIES IN VOGUE</u>
© Conde Nast Publications Limited 1983

SHIRLEY TEMPLE IN BUBBLE GUM. Blossom Company bed doll, she is all cloth, with a hand-painted face, 25" high. Shirley Temple, beloved child star, has been collecting dolls since childhood. She contributed to many trends established during the thirties. Farago Collection.

Charlie Chaplin built his empire on the sympathy he evoked from his portrayal of the comically tragic "Little Tramp."

...a London season stifled with strikes, huge taxation, and old nobility glad of crusts in the country, Paris gay with war profiteers and not much else, and New York a place where millionaires meet to talk of the workhouse.

—THE TWENTIES IN VOGUE
© Conde Nast Publications Limited

CHARLIE CHAPLIN AND THE CATS. Charlie is an all cloth English doll created by Dean's Rag, a company which still exists today. He has a silk-screened face and mohair wig, 13" high. He is accompanied by English Felix Cats, circa 1925, which are patterned after the celebrity cat featured in the "Felix the Cat" cartoons. Farago Collection.

French postcards of Josephine Baker
She inspired Norah Wellings, Lenci and other dollmakers.

Josephine Baker, exotic black dancer, began performing in Paris in 1925, where she was known to wear a snake around her neck and walk her pet panther down the Champs-Elysees.

...a 19-year-old girl from St. Louis called Josephine Baker...For years to come she was to be the most talked about woman in Paris. She was the Ebony Venus, whose frenzied dancing and sinuous, naked beauty gave jaded Parisian audiences a fresh image of sexuality. One critic wrote that 'she gave eroticism style'; another compared her to the black Aphrodite who had haunted Baudelaire.

—TONY ALLAN, THE GLAMOUR YEARS   PARIS 1919 — 40
© 1977 Bison Books Corp.

NORAH WELLINGS THEATER. Features nine dark brown, velvet Norah Wellings boudoir dolls with glass eyes. They surround an E. P. McKillop sculpture of a crocodile, or in this case, a "clock-o-dile" dated 1859. Farago Collection.

Norah Wellings, once a chief designer for the English Chad Valley Company, started her own company in 1926, which lasted until 1960. The dolls pictured here are Bermuda Boys and Girls ranging from 12" to 16" high. The doll in the pastel yellow dress (right of center) is marked Allwin, is 22" high, with a skirt that doubles for a lingerie bag, circa 1929. Most Norah Wellings dolls have a label sewn on the bottom of one of their feet.

The exotic beauty of Josephine Baker may have inspired these black lady dolls dressed in grass skirts.

The Soul has Bandaged moments—
When too appalled to stir—
She feels some ghastly Fright come up
And stop to look at her—

Salute her—with long fingers—
Caress her freezing hair—
Sip, Goblin, from the very lips—
The Lover hovered o'er—

—EMILY DICKINSON

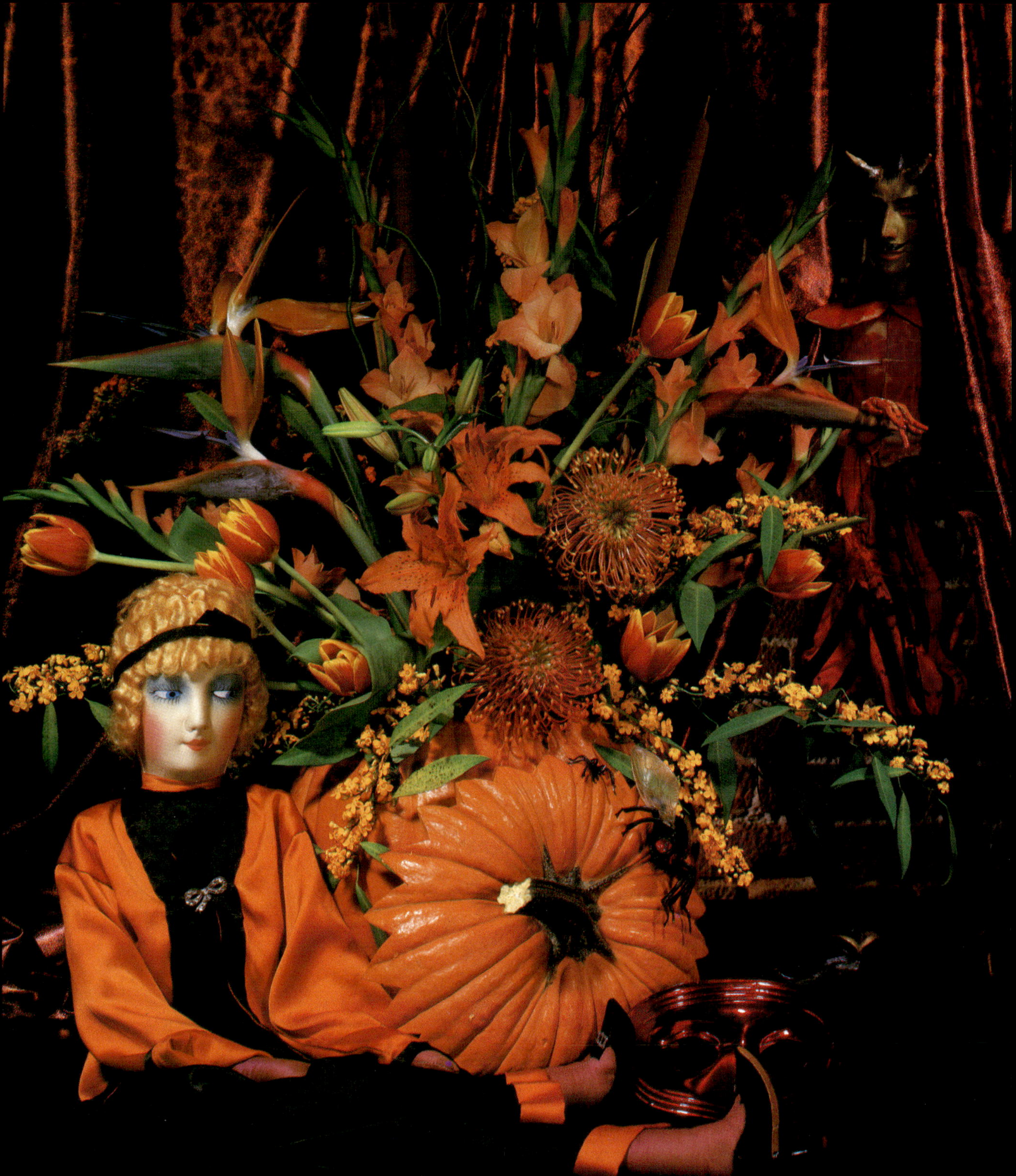

Black, then, was beautiful, fashionable, and sufficiently novel in France to stimulate an audience's curiosity. The popular acceptance of jazz music had turned black culture from an avant-garde concern into a general vogue.

... In Monmartre, the Charleston had taken over the cancan. You crammed into the Beouf sur le Toit to listen to deafening Negro jazz and rub shoulders with the brilliant and the bohemian, or danced on the smallest dance floor in the world at Le Grand Ecart.

—CAROLYN HALL, THE TWENTIES IN VOGUE

© Conde Nast Publications Limited 1983

HAWAIIAN HARVEST. Created by the English cloth doll company, Norah Wellings. It is possible that the female doll was inspired by the exotic beauty of Josephine Baker. They are 36" high, made of dark brown velvet with glass eyes, circa 1929. Farago Collection.

The Soul has moments of Escape—
When bursting all the doors—
She dances like a bomb, abroad,
And swings upon the Hours...
—EMILY DICKINSON

LITTLE EGYPT. Possibly an early Lenci, is all felt, with one trick leg which dangles while the rest of her body is rigid. She wears a rhinestone-studded costume and is 24" high. Farago Collection. The carousel horse, courtesy of Douglas and Jeri Duncan. This particular horse was the first one acquired for their extensive collection of carousel animals.

*Space and time disappear
In the caves of the spirits,
Pain and joy disappear
On the shores of our Fortunate Isles.*
—KIN P'ING MEI

PING PONG. A French, all silk doll, with silk-screened face and printed pattern on body. She wears black silk pajamas and anklets with plastic beads. Could possibly have been inspired by the beauty of actress Anna Mae

A German, musical, composition-headed bed doll with cloth body, 26" high. She is a portrait doll of silent screen star Pola Negri.

When one is in love, one always begins by deceiving others. That is what the world calls romance.
—OSCAR WILDE, THE PICTURE OF DORIAN GRAY

SUCCESS. Three Parisian boudoir dolls, all cloth, 29" high, were made around the turn of the century. A note accompanied the dolls which stated that the groom and the bride's mother were in cahoots to marry her off. Gladyse Hilsdorf Collection, Fayetteville, N.Y.

CADET. (above) An American cadet, has a one-piece composition head, neck and shoulders, with molded and painted hair, composition boots and cloth body, 30" high, circa 1940, manufactured by Freundlich, Germany. Strong Museum, Rochester N.Y.

"OUR LINDY." (right) Charles Lindbergh, American-made by the Regal Doll Company, label attached to suit, sculpted by E. Peruggi, circa 1928. Strong Museum, Rochester N.Y.

TEDDY ROOSEVELT AND THE BEARS. American, papier mâché, with cloth body, 25″ high, sits on a horse with authentic hair (possibly made by Dean's Rag), circa 1903. Gladyse Hilsdorf Collection, Fayetteville, N.Y.

Teddy Roosevelt, 26th president of the United States, author, explorer and hunter, spared the life of a baby bear cub. Soon after, he attended a dinner in his honor, where Steiff bears were placed in the center of each table to romanticize Roosevelt's episode with the bear . . . thus the Teddy Bear got its name.

KING EDWARD AND QUEEN ALEXANDRA OF ENGLAND, (left) after their coronation, circa 1900. Made in England of cast plaster head, hands and legs, with cloth bodies dressed in red velvet and white velour simulating ermine, with gold braiding and trim, 30" high. Gladyse Hilsdorf Collection, Fayetteville, N.Y.

KITSCHY COUPLE. (below) English, silk hand-painted mask faces, both wear synthetic wigs, 26" and are 24" high. The woman is dressed in a taffeta gown and adorned with rhinestones, pearls and large sequins. Strong Museum, Rochester N.Y.

—By my troth
I would not be a queen!
---------------Verily,
I swear, 'tis better to be lowly born,
And range with humble livers in content,
Than to be perk'd up in a glistering grief,
And wear a golden sorrow!
   —SHAKESPEARE, King Henry VIII

WAX PEDDLER DOLL. An English bees-wax doll, with blue glass eyes, circa 1840. Peddler's table is 18" across and seated figure is 10" high. She is a fragile and detailed study. Props include gloves, goblets, shoes, samplers, birds, pastel burners, beads, scissors, thread, knitting and everything one might find at their local flea market. Ruth Noden Collection.

Noble blood is an accident of fortune; noble actions characterize the great. —GOLDONI

PRINCE RAINIER AND PRINCESS GRACE OF MONACO by Martha Thompson, American NIADA artist, 1957. Bisque shoulder-head, lower arms, hands, lower legs and shoes, molded hair with cloth bodies. Both with sculptured, painted features, they wear replicas of the clothes worn on their wedding day. Grace's dress is lace and is dotted with pearls; she carries a lace and pearl covered prayer book and wears a blue garter. Musée National de Monaco, Collection de Galéa.

GEORGE AND MARTHA WASHINGTON by American dollmaker Emma Clear. Bisque heads, molded hair, neck, shoulders, lower arms and legs, with cloth bodies, 30" high. Emma Clear made beautiful bisque dolls reminiscent of older dolls. In this case, these dolls are totally original in style and concept. Sculpturally, they are exquisite—especially the heads and hands. Gladuse Hilsdorf Collection, Fayetteville, N.Y.

> The sunrise wakes the lark to sing,
> The moon rise wakes the nightingale
> Come, darkness, moonrise, everything
> That is so silent, sweet, and pale:
> Come, so ye wake the nightingale.
> —CHRISTINA ROSSETT

VALENTINE. (above) Two Anita dolls: the doll (left) is dressed in a silk Apache costume, and the other has yellow gold silk braided hair and a satin pant suit, 31" high. Sold without clothes, these dolls exhibit work well done by talented collectors. Farago Collection.

> A sympathetic heart is like a spring of pure water bursting forth from the mountainside.
> —ANONYMOUS

ROSEBUD. (below) A French boudoir doll, circa 1925, by Bechoft, 12" high. John Darcy Noble Collection.

PEACOCK PIERRETTE. *She wears the typical black and white costume of Pierrette and is probably a European boudoir doll. All cloth with a hand-painted chamois face and mohair wig, 28" high. Some say she resembles Jean Harlow, others say Marlene Dietrich. Tony Menchin Collection.*

LATIN LADY. (left) Possibly a South American, Lenci type, hand-painted felt face and felt body, sewn-in black silk hair, with bakelite earrings, dressed in organdy with felt flowers, 29″ high. Richard Wright Collection.

IVORY SISTERS. (below) Possibly Portuguese or French, with ivory heads and hands, dressed in velvet with metallic trim, were probably crêche figures made between 1776 — 1870 and may have been used as telephone covers in the early 1900's. Strong Museum, Rochester N.Y.

CRITICS. American "big faced" (largest of the heads made) boudoir dolls engaged in the critical viewing of <u>Romeo and Juliet</u>. Their heads are hand-painted composition. The one on the left has an all cloth body, while the one on the right has composition lower arms, hands, lower legs and feet, 37" high. Farago Collection.

O love's best habit is a soothing tongue.
—SHAKESPEARE, Romeo and Juliet

> Every king springs from a race of slaves, and Every slave had kings among his ancestors.
>
> —PLATO

OTHELLO. A German doll, with black ebony carved head and hands, and inset ivory teeth. Made in the 1890. 20" high. Gladyse Hilsdorf Collection, Fayetteville, N.Y.

The city's reputation for licentiousness was such that in bars and clubs across America the very mention of its name was enough to produce a knowing wink and a leer. Books were written about it, films were set there, and songs were sung, "How ya gonna keep 'em down on the farm after they've seen Paree..?

...The mood of the day was extravagant and eccentricities flourished. While the Dadaists and their Surrealist offspring were inspiring riots, the Duchess von Fretag-Loringhoven was to be seen on the terrace of the Deux Magots in a hat decorated with a large watch and chain.

—TONY ALLAN, PARIS IN THE GLAMOUR YEARS

THE SURREALIST CAFÉ. A European male doll, dressed as an Apache dancer, with hand-painted felt face, black felt hair, muslin body stuffed with excelsior; he is not jointed and stands 26" high. He is complete with dangling cigarette and roguish attitude, as he earnestly pursues his friends, the Lenci Fadettes, 26" high (as seen on pages 122 and 123). Lenci boudoir doll plays the piano, height 30" (same face as the Dietrich doll page 129). Farago Collection.

This set was created to simulate a cafe on the left bank during the Roaring Twenties. The walls are papered with the French newspaper, Le Figaro, the paintings are by Salvador Dali, Magritte, Renoir, Mucha, Manet, di Chirico and Vermeer.

. . . you are leaving me in the most beautiful mood of my life, in the phase of my love that is most real, most passionate, and most replete with suffering! . . .

—GEORGE SAND

(George Sand's granddaughter, Lauth Sand was a dollmaker)

REGRETFULLY, CLARK. *She is German and marked "modele depose" with the outline of a boot inscribed on the bottom of her leather shoes. She has a hand-painted silk face, an all cloth body, with legs jointed at the hip. She wears felt clothing with ornately embroidered garters and undergarments. 24" high. Farago Collection.*

In everything there is an unexplored element because we are prone by habit to use our eyes only in combination with the memory of what others before us have thought about the thing we are looking at. The most insignificant thing contains some little unknown element. We must find it.

—MAUPASSANT, Preface to Pierre et Jean, 1887

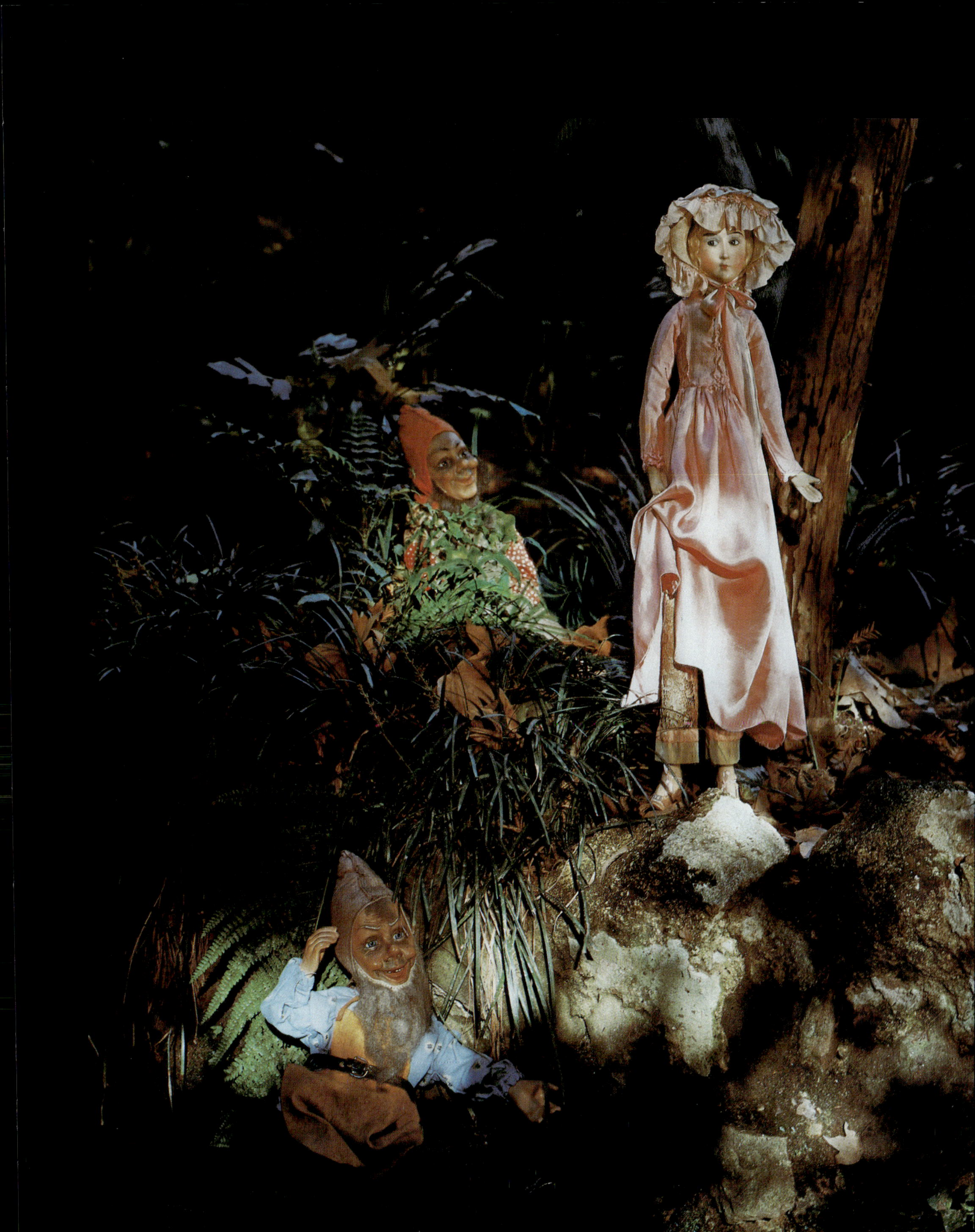

AGATHA CHRISTIE'S INSPIRATION. During the 1920's Agatha Christie was a client of the high-society dressmaker, Nora Crampton. Upon seeing this large cloth doll exhibited in Crampton's window, Christie wrote a series for a magazine called the "Dressmakers Doll." The doll is 48" high, all cloth, with hand-painted face and hair.
Betty Titone Collection.

ATTIC SUITE. Two American all-cloth dolls with hand-painted faces, 30" high. They reminisce with Deco greeting cards. Farago Collection.

Love is the marrow of friendship, and letters are the elixir of love. —JAMES HOWELL

BETWEEN TAKES. Almost real, these two boudoir beauties strongly resemble Hedy Lamarr. Doll masks such as these have been found in Barcelona, where there might have been a factory. They are definitely European, with stockinette-faces, silk thread hair, eye lashes, 30″ high. They wear velvet & taffeta dresses. Farago Collection.

> No one should die without seeing himself in the flesh on the 'Movie' screen. It is the oddest sensation to look calmly on at one's other self walking, smiling, dancing, talking and, sometimes, doing the most ridiculous things. It is not always flattering to one's vanity, but very amusing to one's friends.
>
> —CAROLYN HALL, THE TWENTIES IN VOGUE
> © Conde Nast Publications Limited 1983

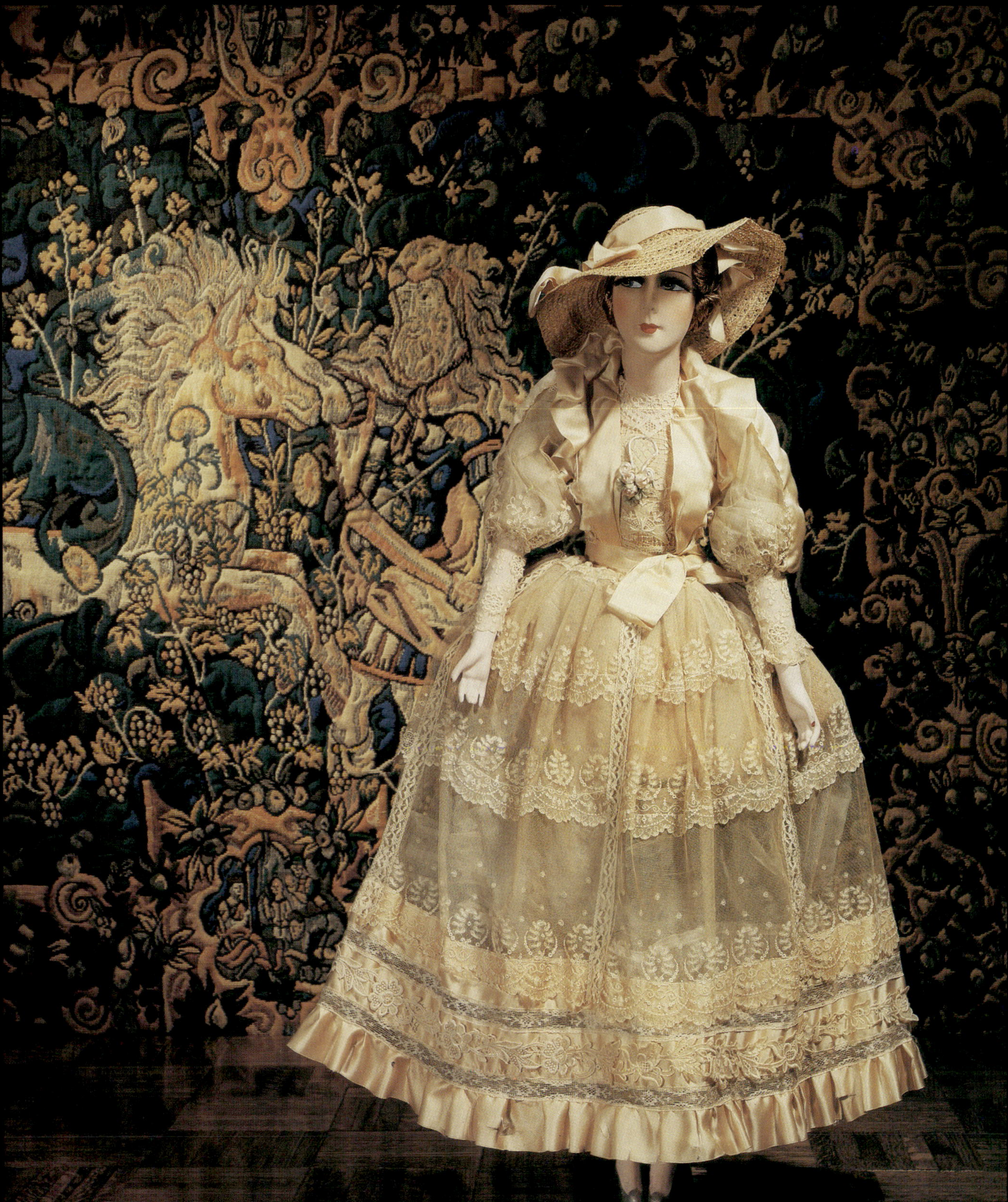

*Whoever undertakes to set himself
up as judge in the field of Truth and Knowledge,
is shipwrecked by the laughter of the Gods.*

—ALBERT EINSTEIN

HARRY LAUDER AND FRIEND. *Famous Scots minstrel Harry Lauder, known for his magnetism and songs like "Quit Your Ticklin' Jock," sits on the waterfront with a shipment of antique advertising tins. His face is papier mâché. His friend may have been used in a Western film as a portrait doll of one of the stars. She is made of papier mâché and stands 32" high. Farago Collection. Tins: Nicholas Farago Collection.*

"Evil is not destructive to the good alone but inevitably destroys itself as well. For evil, which lives solely by negation, cannot continue to exist on its own strength alone...

In the end evil perishes of its own darkness, for evil must itself fall at the very moment when it has wholly overcome the good, and thus consumed the energy to which it owed its duration.

—THE I CHING

DEVIL. A German papier mâché doll, 18" high, with a cloth body, said to be circa 1900 — 20. Oddly, it also has red rubber hands. Gladuse Hilsdorf Collection, Fayetteville, N.Y.

Although your wicked brows belie
   the angel in your eyes,
it is a blessed sorcery
   by which I am beguiled:

with all the ineffectual awe
   of prostrate votaries
I worship at your trivial
   and tantalizing shrine!...
   —CHARLES BAUDELAIRE,
      "Song For Late In The Day"

THE WITCH OF KLIMT. She is an Anita bed doll, wearing one of the most elaborate costumes ever seen on one of these dolls. All in tatters, resembling Dicken's Miss Havisham, she retains a luminescent and bewitching beauty. The backdrop was inspired by Gustav Klimt. Farago Collection.

"To catch Dame Fortune's golden smile,
    Assiduous wait upon her;
And gather gear by ev'ry wile
    That's justified by honour;
Not for to hide it in a hedge
    Not for a train-attendant,
But for the glorious privilege
    Of being independent."
—ROBERT BURNS

INDEPENDENCE EVE. She is a French boudoir doll, with hand-painted stockinette face, long lashes, silk thread hair, plaster lower arms, hands, lower legs and high-heeled shoes (marked Paris on the sole) wearing a blue silk file dress and maribou feathers, 30" high. Farago Collection.

We are but older children, dear,
Who fret to find our bedtime near.
　　—LEWIS CARROLL, <u>ALICE IN WONDERLAND</u>

GOLDEN POND REVISITED. All cloth European bed dolls are hand-painted mask faces, 27" high. Definitely manufactured and not one-of-a-kind, many faces like these have been located in Barcelona, which indicates a factory may have existed there. Farago Collection.

Bronze plaque by Riva, (above), 1915, commemorated the marriage of Elena and Enrico Scavini. Carla Caso Collection. Photograph by John Axe.

Elena Koenig di Scavini (below), 1915, photographed by Enrico Scavini. Photocopy by John Axe.

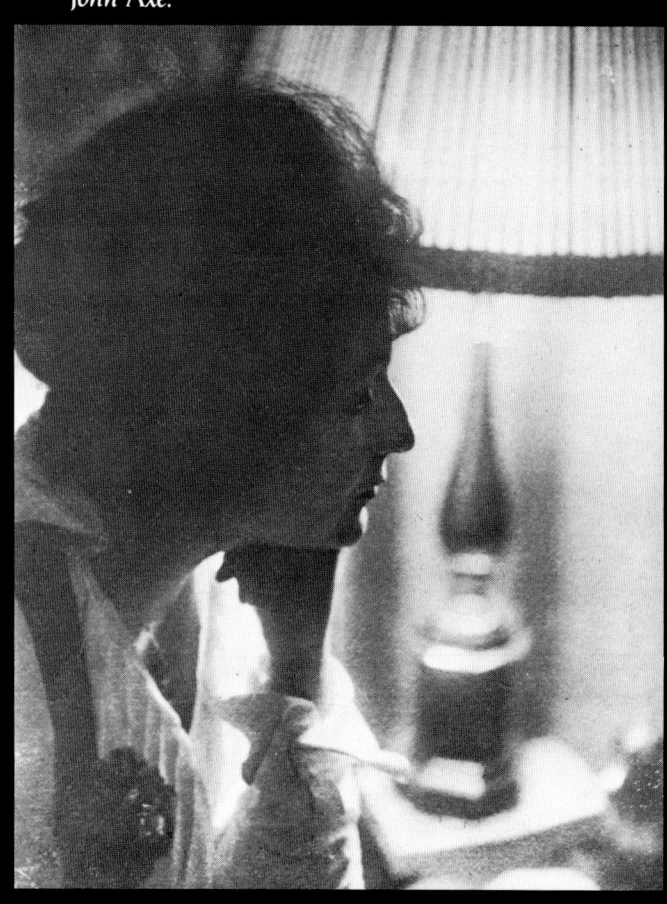

III. Lenci, Virtuoso of the Boudoir Doll

"Mimi," a charming Lenci boudoir doll, invites you to an enchanting visual feast.

Boudoir dolls, as seen in the 1931 Lenci Catalogue.

The Lenci Company is one of the only early doll companies which still exists today. Under the directorship of Beppe Garella, fine felt dolls for children are still being made. The original Lenci Company was founded by Enrico and Elena di Scavini in 1919. Their daughter, Anili, also continues the tradition of fine felt dollmaking for children at her company called Anili.

The genius who inspired one of the most incredible doll empires ever known was Helenchen (Elena) Konig di Scavini, known to many as "Madame Lenci." Her personal friends ranged from Josephine Baker to Mussolini. She was a chain smoker, whose subject matter was intrinsically entwined with the magic of movies. Although movies were silent and black and white at the time, the Lenci dolls were always a rapture of color and design. The Scavinis assembled a company of nearly 1,000 artists, illustrators and craftspeople, who were among the most brilliant in Italy and Europe.

The family reports that Helenchen, a small German girl, and her sister ran away to the Circus of Madame Nouma Hawa. Eventually they returned home from this difficult adventure to the comfort of their home. Yet, Helenchen was ingrained with a feeling of wanderlust that was not satisfied by her experience with the circus. Eventually her abundance of creative energy led her to courageously open her own portrait studio in 1909, after she received her Master's Degree in photography. Her nude photographic subject matter reveals that she was

Art photograph, 1914, signed "Lenci," taken by Elena prior to her marriage to Enrico. Carla Caso Collection, Photocopy by John Axe.

Art Dolls

brave before it was fashionable. She also worked in textiles, including embroidery and "batica"—wood stamping of patterns on cloth.

Enrico di Scavini, Elena's sensitive, Italian suitor, courted her for four years before they married. She finally consented to marry Enrico when she had succeeded in launching her photography business, and he had become a self-sufficient businessman. They married on June 7, 1915 and settled in Torino, Italy. Enrico went to war in 1917. When her infant daughter died, Elena was overcome with grief and loneliness. Being a gregarious and resilient woman, she set to work creating dolls to fill up her empty hours waiting for Enrico to return home. Enrico was so delighted with the dolls she had made, that he actually tried to find a market for them in Turin, but he was told the happy rag dolls were too simple. The Scavini's were soon looking forward to being parents again, until Elena miscarried. Once again, despondent with grief, she began to create a more sophisticated type of doll. A photograph of her in her 1915 studio in Dusseldorf shows a Steiff

Deco lady (above) featured on Italian postcard, appears with her Lenci doll, a mascot in her private parlor.

Boudoir dolls (below), 28" high, resemble Marlene Dietrich and Louise Brooks. As seen in 1930 Lenci Catalog.

Art Dolls

Lenci permanent showroom, features Rudolph Valentino and boudoir dolls in foreground. Playthings, March 1928.

Mama Katz character doll on her work table. Obviously, the Steiff dolls were one of her earliest influences.

Although she kept her first creations in a closet, one evening they were brought before an entrepreneurial friend who insisted on taking them back to the States, where he was sure they would sell. His hunch was right. Americans loved the Lenci dolls and became the Scavini's first patrons. Ironically, the sales of the first Lenci dolls took place aboard ship, long before reaching American soil. The American marketplace for Lenci dolls stimulated the beginning of one of the most creative and innovative ventures in doll history.

After sending to Borsalino for the strong wool felt usually used in blocking hats, the Scavinis began production. They became so successful they had to hire a staff to fill the surprising number of orders. They named the company L.E.N.C.I., which stands for the Latin phrase "Ludus Est Nobus Constanter Industria"... which translates to "To us, play is a constant industry."

The dolls were designed by a number of artists, however, the Lenci tag was their only

develop artistic tendencies in your children
Buy a Lenci Doll
- to place on hassocks
- to lend color to boudoir.
- to decorate the corner of your limousine.

Playthings magazine, July 1923.

Elena Scavini with young girls on tour of the Lenci Company. Shirley Buckholtz Collection.

identifying mark. The Lenci company also produced sensuous ceramic sculptures. Sculptor Sandro Vacchetti, a close friend of Elena di Scavini's, became enchanted by her first dolls. In 1919, along with his brother Emilio, Vachetti joined her in her workshop. Soon after, the famous illustrator Marcello Dudovich was employed to do preparatory drawings for the dolls, and Giovanni Riva modeled heads for the up-and-coming company. Eventually Vacchetti became art director of the company until 1934 when he started his own china business called Essevi. Other sculptors involved in china (and possibly dollmaking) were Nillo Beltrami, Clelia Bertetti, Lino Berzoini, Gigi Chessa, Guilio Da Milano, Teonesto De Abate, Claudia Formica, Grande, Beppe Porcheddu, Massimo Quaglino, Mario Sturani and Felice Tosalli.

The Scavinis and their marvelous staff had impeccable taste and integrity in the production of their products. They were imitated constantly. They ran a multifaceted company which was as perfectionistic at sculpture as it was with dolls, mannequins, purses, clothing, cradles, furniture, advertising and photography. All of their work was flawless and amazing.

Elena di Scavini was part of the driving impetus behind the dynamic sculptural events that gave birth to the Art Deco world of the 1920's and 1930's. Around 1926 – 28, Walt Disney asked her to work with him, but she chose not to since her own company was in its heyday at that time. Many of the Lenci ceramics exhibit the look that later appeared in Disney's film *Fantasia*.

The Scavini's, above all, had incredible vision. There was never an uninteresting or mediocre item produced. Every article bearing the tag *Lenci di Scavini* was truly a work of art. In the following pages you will delight in the charming men, romantic women and zany creatures that were among the first signs of Art Deco sculptural fantasy.

Playthings magazine, January 1923, shows Opium Smoker at right.

Lenci
DI
E. SCAVINI
58 WEST 45TH STREET
NEW-YORK CITY

Hairnet package illustrates the recurring theme of Pompadour and Pierrot.

They may talk of love in a cottage,
    And bowers of trellised vine—
Of nature bewitchingly simple,
    And milkmaids half divine . . .
But give me a sly flirtation,
    By the light of a chandelier—
With music to play in the pauses,
    And nobody very near.

. . . True love is at home on a carpet,
    And mightily likes his ease—
And true love has an eye for a dinner,
    And starves beneath shady trees.
His wing is the fan of a lady,
    His foot's an invisible thing.
And his arrow is tipp'd with a jewel,
    And shot from a silver string.

—N.P. WILLIS "Love in a Cottage"

POMPADOUR AND PIERROT. *Flirtatious mistress of Louis XVI tantalizes the loverlorn Pierrot. All felt dolls, 26" high. Pompadour wears pastel pink organdy dress with felt rosettes. Pierrot wears all felt pant suit with large, flat felt buttons. Farago Collection.*

All the delight having perished,
hopeless I remain:
It was only a dream of Spring!
—UNKNOWN JAPANESE WOMAN

MADAME BUTTERFLY. A stunning Geisha reflects in a fantasy pool with water lillies. All felt with silk thread embroidery, rooted black mohair wig, 26″ high. Farago Collection.

SHOPPING SPREE. Features twin "Mimi" dolls shopping at an Art Deco marketplace. Real miniature fruit and vegetables were selected to fit the scale of the dolls. All felt dolls with organdy bloomers adorned with flowered garters, 26" high, circa 1927. Farago Collection.

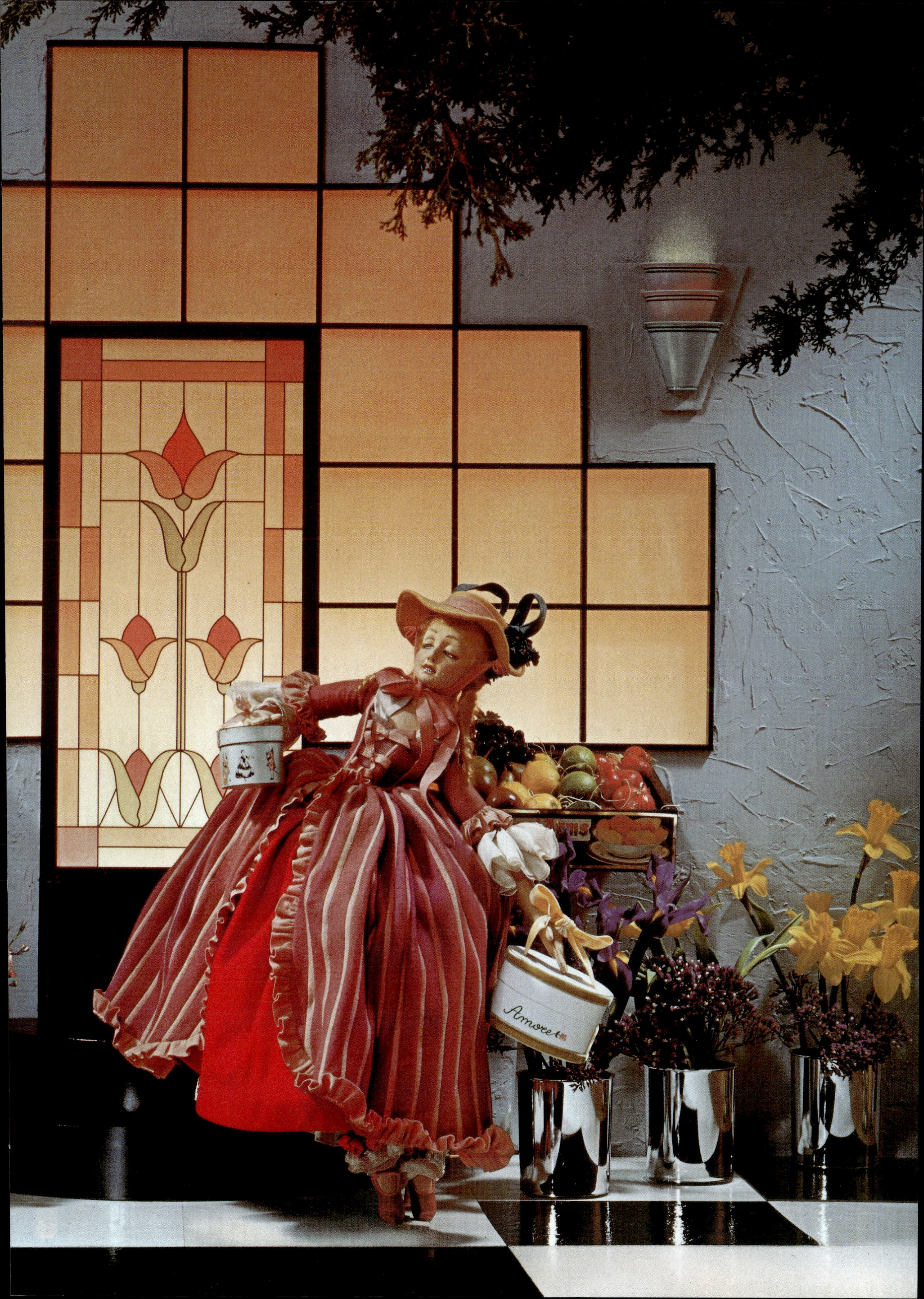
Amore

MARILYN MILLER. All felt Lenci lady with ruffled organdy dress and white fur cat on her shoulder represents Ziegfeld girl, Marilyn Miller. She was famous for her rendition of "Look For The Silver Lining." Museum of the City of New York.

JUNE BRIDE. *Wistfully waiting under a willow, she placidly ponders her future in solitude. Her heavy eyelids are reminiscent of Marlene Dietrich. All felt with organdy dress and felt flowers, she is constructed to sit. Collection of Countess Marie Tarnowska, London.*

> Oh Life! accept me—make me worthy—teach me.
> I write that. I look up. The leaves move in the garden,
> the sky is pale, and I catch myself weeping...
> —KATHERINE MANSFIELD

Vanity Fair, 1922, advertised and offered this doll for sale in their Christmas gift article. They called her a French doll, however, she is unmistakably an early Lenci Fadette.

"It was said that the craze for the Lenci dolls was not limited to women. He found that many men were among the purchasers generally favoring the sophisticated French chorus girl, legs crossed, with a cigarette tilted at an angle."
—Toys and Novelties. July, 1923.

Men always want to be a woman's first love. That is their clumsy vanity. We women have more subtle instinct about things. What we like is to be man's last romance.
—OSCAR WILDE, A Woman of No Importance

FADETTE. *Sassy smoking flapper sits atop carved wood tramp art box. She is all felt, with a rooted mohair wig, leather high heels, 26" high, circa 1922. Ferago Collection.*

SPANISH LADY MANTILLA *has an exceptionally haughty and exotic look. She is adorned with beautifully detailed hand-painted wooden fan with a scene of a bull and fighter. She also wears a hand-painted, wooden hair piece as well as a black felt headdress. Farago Collection.*

Preliminary rendering of a Spanish dancer by Deco period Hollywood film poster artist Marcello Dudovich. Courtesy of Lenci Garella.

SPANISH SERENADE. *She is an all felt Spanish lady, 26" high, with a silk thread embroidered shawl, backlit by a glowing, golden orange tree. Herb Scott Collection.*

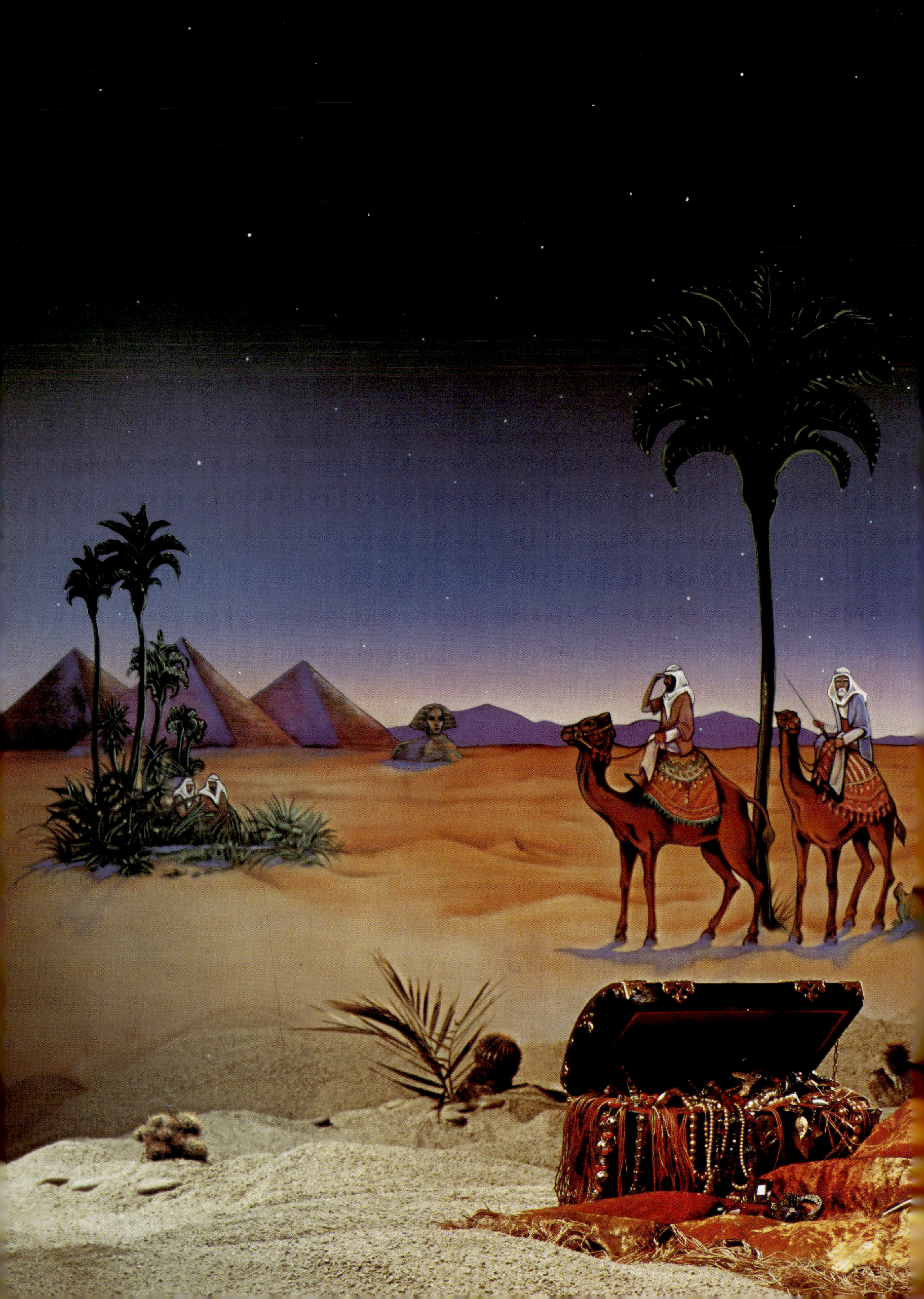

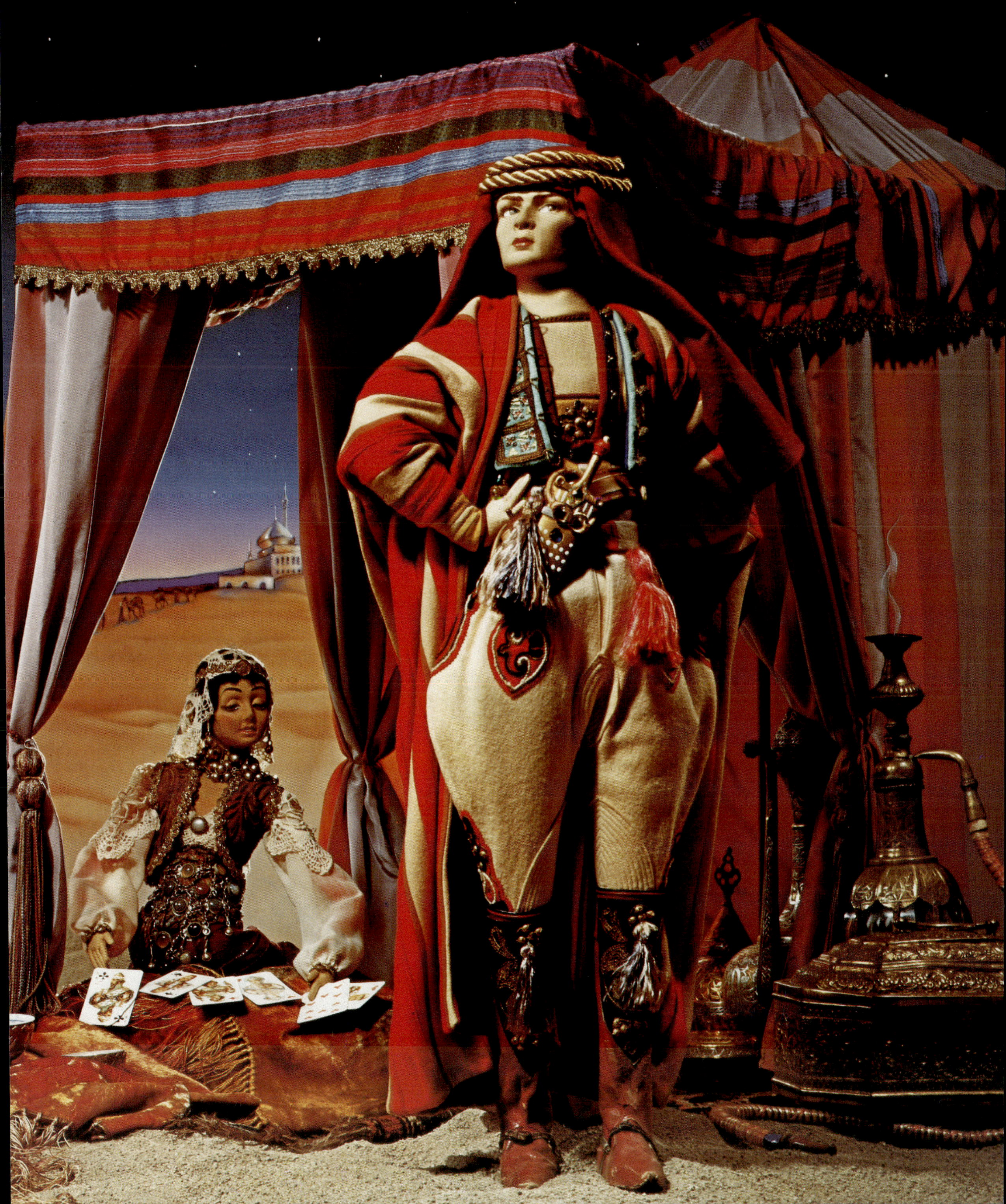

Lenci's Marlene Dietrich doll, Ruth Noden Collection.

Marlene Dietrich poses with her Oriental Lenci doll, 300 series.

Photo courtesy Richard Rheem

THOSE HOT DESERT NIGHTS. *(previous page)* The veil of night falls as Rudolph Valentino listens skeptically to his fortune-telling lover. The Valentino doll is the ultimate graven image. This fabulous Lenci creation was designed to promote the film "Son of the Sheik"(circa 1927). He is all felt, 30" high, and all original, except for the gold braid around his head. His leather boots are embroidered with metallic threads, adorned with silk tassles and spurs. He carries a wooden dagger dotted with glass stones. His lover was a delapidated Spanish dancer we refurbished and dressed as a harem girl, 26" high. Both dolls:

MARLENE DIETRICH. *This incredibly realistic doll is called "Marlene Dietrich," however, many other Lenci dolls resemble her far more than the one pictured here. The features of this doll are not as classically glamourous as they are refreshingly human. She is all felt, 30" high. Ruth Noden Collection.*

Every day the women of today
Are up to new and better tricks.
They even play the banjo now,
Drink cocktails, drive their own cars.
Oh, me! Oh my!
Well, now they're going themselves one better:
It wasn't enough to cut off their hair,
Now they're letting us see their calves
And even higher, all the way up!

—Chanson de Geste, (Old French Epic Poem)

LIBERTINA. *She has the same face as the "Dietrich" doll (previous page) and was sold with the chair. She is all felt, with velvet robe appliquéd with felt designs, 30" high. Farago Collection.*

There is only one man in the world
and his name is All Men.
There is only one woman in the world
and her name is All Women.
There is only one child in the world
and the child's name is All Children.
—CARL SANDBERG, THE FAMILY OF MAN

Young brown girl (center) wears a natural grass skirt and wooden hand-carved and painted elephant on her necklace, 15" high.

Salome (bottom right) bares a striking resemblance to Josephine Baker, however, this doll appeared long before Josephine Baker made her debut as an exotic black dancer in the Paris Folies-Bergère. Elena Scavini and Josephine Baker later became friends, and a doll representing her was eventually sculpted. Often the Lenci dolls predicted fashion as well as set trends for the up and coming stars and starlets. She is 18" high, 1922.

A 12" black boy sits atop a rare jointed and airbrushed wooden Lenci tiger that wiggles when its tail is moved

FAMILY TREE. A hand-carved, Balinese banana tree provides a perch for four very rare, early black Lenci dolls. Black baby (top) is wearing wooden hand-carved and painted jewelry—a white tooth earring and a serpent necklace along with colorfully embroidered diaper and hat, 11" high. All dolls Farago Collection.

FUKURUKO. *Japanese God of Wisdom and Longevity, 12" high. Farago Collection.*

LENCI ECCENTRICS *mindlessly ignore impending danger of an active volcano. Features Fukuruko, 11" high, Chinese man, 14" high and Opium Smoker, 11" high. Actually, they probably are all high. Farago Collection.*

"A man depends largely on the woman for the light in the family... With a few words, a woman can give meaning to a whole day's struggle, and a man will be very grateful."

—ROBERT A. JOHNSON, SHE

A PLEASANT PEASANT. A 30", rare-faced, all felt Lenci, dressed in detailed Russian costume, whose charming smile warms up a cold wintery scene. Richard Wright Collection.

COUNTRY GIRL. Spinning wool while she sits amidst miniature barnyard scene, she is all felt with organdy scarf and headdress, 25" high. Ruth Noden Collection.

NOT A CREATURE WAS STIRRING, *except for this startled and resplendent Russian Lenci girl. She is all felt with multicolored streamers adorning her blonde mohair braids. 25" high. Ruth Noden Collection.*

*And out of the frozen mist the snow
In wavering flakes begins to glow;
    Flake after flake
They sink in the dark and silent lake.*
—BRYANT, "The Snow-Shower"

CHRISTMAS EVE. Dashing home with packages and her diminutive dogs, this rare Pierrette doll wears an elegant tiered skirt. She is also pictured on page 40 with Claire Windsor. She is all felt with black and white yarn tassles on shoes, elbows and hat, with black felt flounce and cuffs. 25" high. Farago Collection.

may i feel said he
(i'll squeal said she
just once said he)
it's fun said she

(may i touch said he
how much said she
a lot said he)
why not said she

(let's go said he
not too far said she
what's too far said he
where you are said she) . . .
—e.e. cummings

(Mozart was well known for his scatalogical playfulness.)

MOZART AND MARIA THERESA, *posed before their coach. These dolls were found in Argentina and wear very unusual clothing for Lenci dolls. The clothing is composed of silk brocades, metallic trims, velvet bows and rhinestone accessories, 25" high. Farago Collection.*

At Christmas I no more desire a rose
Than wish a snow in May's new fangled mirth;
But like of each thing that in season grows.
—SHAKESPEARE, Love's Labour's Lost

Barbier Snowscape. George Barbier painted this illustration for "Outfits for St. Moritz" which appeared in the <u>Journal des Dames et des Modes</u> in February 1913.

From the GOLDEN AGE OF STYLE by Julian Robinson, ©1976 by Orbis Publishing Ltd. Reproduced by permission of Harcourt Brace Jovanovich, Inc.

SONIA STOLICHNAYA in *Barbier Snowscape*. We re-dressed this 25" all felt Lenci in a Deco costume after the Barbier design. It is a challenge to re-dress an old doll with appropriate and tasteful attire. This costume was implemented by designer Ricardo Figuilera. Farago Collection.

KAZATSKY DANCER. A redhaired Russian boy leaps into the air, showing off for his amused audience nestled in a pastoral scene. Russian boy, 35″ high, is one of the most limber of all Lenci dolls, and assumes many joyous poses. All felt, he wears green leather boots embroidered with yarn.

Russian woman (standing next to wheel) has a very rare and realistic face, wears a felt and organdy costume with felt-flowered headdress, 26″.

Girl sitting on wagon (far right), wears a large, brilliantly colored headdress and costume of felt and organdy, 35″ high.

Russian children, 18″ high, wear duplicate costumes of the larger dolls. All dolls: Farago Collection.

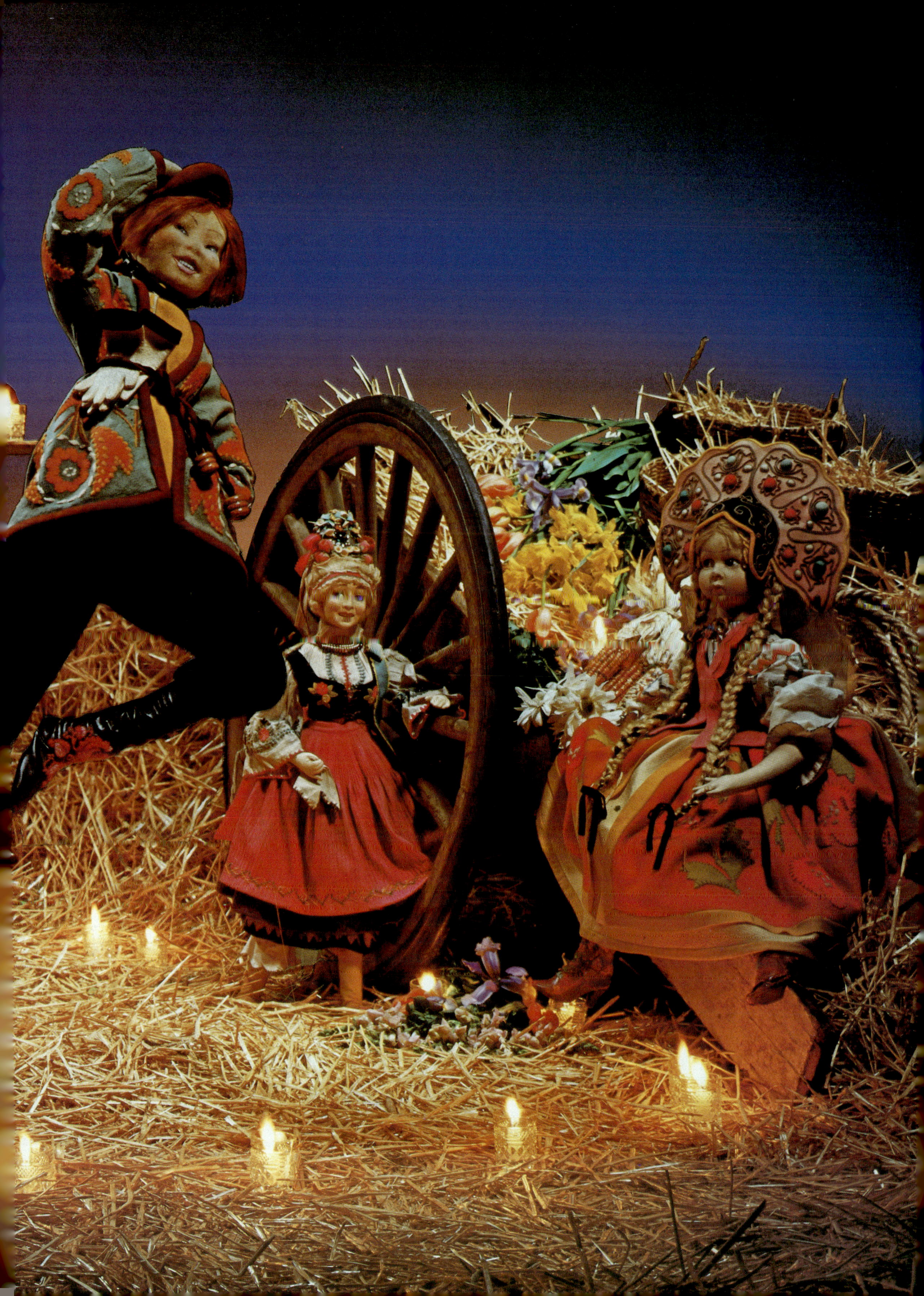

CENTAUR AND FAUN (left). 14½" high. this extraordinarily sensuous sculpture was created for the Lenci Company by Tosali, who was known for his expertise with animals. The male figure bears a striking resemblance to Tosali and may be a self-portrait. Farago Collection.

MIDNIGHT LADY (below). Sculpted by Sandro Vachetti, one of the first designers to work with Elena Scavini. He later became art director of the company while he continued to design dolls and ceramics. This sculpture is marked Essevi, the name of his independent ceramic company. 9½" high. Farago Collection.

TURTLE LOVE. This lovely marine mother was designed before Disney's film, <u>Fantasia.</u> It is known that Walt Disney asked Elena Scavini to work with him, but she declined since her business was flourishing at the time. Possibly sculptures such as these inspired his amazing cartoon fantasy. 16" high. Farago Collection.

The arrival of a good clown is worth 20 asses laden with drugs.
—AGE-OLD QUOTE FOUND BY NORMAN COUSINS

IV. Bring In The Clowns

The Banana Man presents a parade of Pierrots, Harlequins, jesters and clowns.

Art Dolls

Pan, the Greek God of nature, woods, fields and flocks established the tradition of the community minstrel and entertainer. The nymph Syrinx fled from him and was transformed into a bed of reeds. He took reeds of unequal length and invented the syrinx, a shepherd's pipe.

Laughter is nature's best pain killer. Pierrots, harlequins and clowns with their twinkling side-glancing eyes, sardonic smiles and mischievous expressions are the stand-up comedians of the doll world. Pan, Puck, Punch, Puchinello, Polichinelle and Pierrot represent the essence of mankind at his best and his unfortunate worst. The word Pan, from Greek religion, also means "all," possibly meaning all men. "The Fool," who knows everything, knows nothing, represents "Everyman" and is depicted in the Tarot cards of the ancient Jewish Cabala. Along with the jester, harlequin and clown, these characters share in the design to entertain humanity and are the basis for all comedy.

At first, only royalty could afford the company of a hired comedian to smooth over court intensities. During the Renaissance in Italy, improvisational acting troups evolved from the King's court into the street, where they could earn their living performing for audiences as they travelled. They called themselves the *Commedia dell' Arte* — which translates to *Comedy of Skill*. The actors were expected to sing, act, do acrobatics and dance. One of the most important functions of the Commedia was "to let off steam."

Early Pierrot costume.

What had been totally accepted in the Middle Ages was now open to philosophical question. Renaissance men were full of

Commedia dell' Arte, a porchoir print by George Barbier, shows Harlequin, Pierrot and Columbine playing before a conductor and orchestra under the stars as it actually may have been.

Columbine and Harlequin in Le Carnaval. *Dancers Lopokova and Idzikowski.*

Courtesy Nesta Macdonald.

vitality, criticism and satire...ready to conquer life while constantly examining human weakness. To give a poor performance was tantamount to a betrayal of all the Renaissance stood for and the Commedia was part of the fine art movement at that time.

The roots of the Commedia and its characters have been traced to the 6th century A.C. improvisational theatre, a crude form of farce and mime which existed in Greece and Rome. When Roman civilization collapsed, the ancient form of comedy was revived in the puppet form with the emergence of the English *Punch*, the Italian *Pulchinella* and the French *Polichinelle*, a grotesque humpback with a hook nose and big belly, he was known to be a boasting, sly, but stupid coward.

Because of religious associations, Punch's morality plays always had the same ending; he beat the Devil. Though he was much like the Devil himself, he somehow managed the conquest (except, perhaps, of the Devil within). By Medieval times, Punch's humor had become insidiously profane, mocking "morality, marriage, paternity, friendship, learning, law, order, death and the Devil himself... [treating] life as an empty joke." Nevertheless, he found his way into the Mystery Plays of the Church until the 15th century, when he was eventually banned.

Puck, also known as Robin Goodfellow, was a small hobgoblin who won literary recognition in Shakespeare's *Midsummer Night's Dream*. A mischievious character addicted to practical jokes, he delighted in the follies of lovers in love. Puck came from a prehistoric fertility ritual and, with the onset of Christianity, became a protagonist in morality plays, where he initiated most of the confusion the plots revolved around.

Pierrot is the culmination of all ancient comedic heritage. He was a buffoon who evolved from the coarse and agile Harlequin character. He was less acrobatic than Harlequin, incorporated more mime and had a naive outspokenness on matters of the heart. In France, when the Italian Commedia dell' Arte came to appear in the court of Henri IV, in 1599, he was not yet called Pierrot; he was probably known as Pedrolino until the 1890's.

Art Dolls

Pagliacci, the clown/fool featured in Ruggiero Leoncavallo's opera.

Truly the innocent fool, Pierrot was hopelessly in love with Columbine in most Commedia plays. He was a tragic yet comedic character who, although simple, became quite philosophical when unlucky in love. He was the mime figure who creatively inspired writers, painters, choreographers, actors, and dancers for centuries. These included Caruso singing *Pagliaci*, Nijinsky dancing in *Petrouchka*, Watteau's paintings of Pierrot, Duberau and Jean Louis-Berault starring in the French 1947 film, *Le Infant d' Paridis*, Charlie Chaplin as the *Little Tramp*, and the inimitable Marcel Marceau, as "Bip," who continues to pantomime on stages throughout the world.

Probably the most interesting portrayal of the fool is the role of the Jester in Shakespeare's *King Lear*. All the time that Lear is losing his mind and power, his Jester is probably the only one who understands him, gives him good advice, can inspire his trust and assuage some of the King's fear with his humor. It is the combination of the serious tragedy, coupled with contrasting humor, that creates dramatic tension in most great plays and plots.

Petrouchka, a clown/fool as portrayed by Nijinsky for the Ballets Russes.

Joe E. Brown, who had a tragic youth, made a fortune making people laugh. Here he wears a harlequin costume.

Al Jolson, who was subject to brooding moods, is dressed as Pierrot. He was one of the best known minstrels of the Twenties. International Museum of Photography, George Eastman House 1982.

Most tragic heroes usually take themselves too seriously. It is the fool, throughout literary and theatrical history, from the court jester in Shakespeare's *King Lear* to Hollywood's Charlie Chaplin, who reminds us there is a practical advantage to laughing at oneself... and that usually things could be worse.

Enough serious talk. For some comic relief proceed to the following page.

A *merry heart maketh like a good medicine.*
—THE BIBLE

MUSEUM

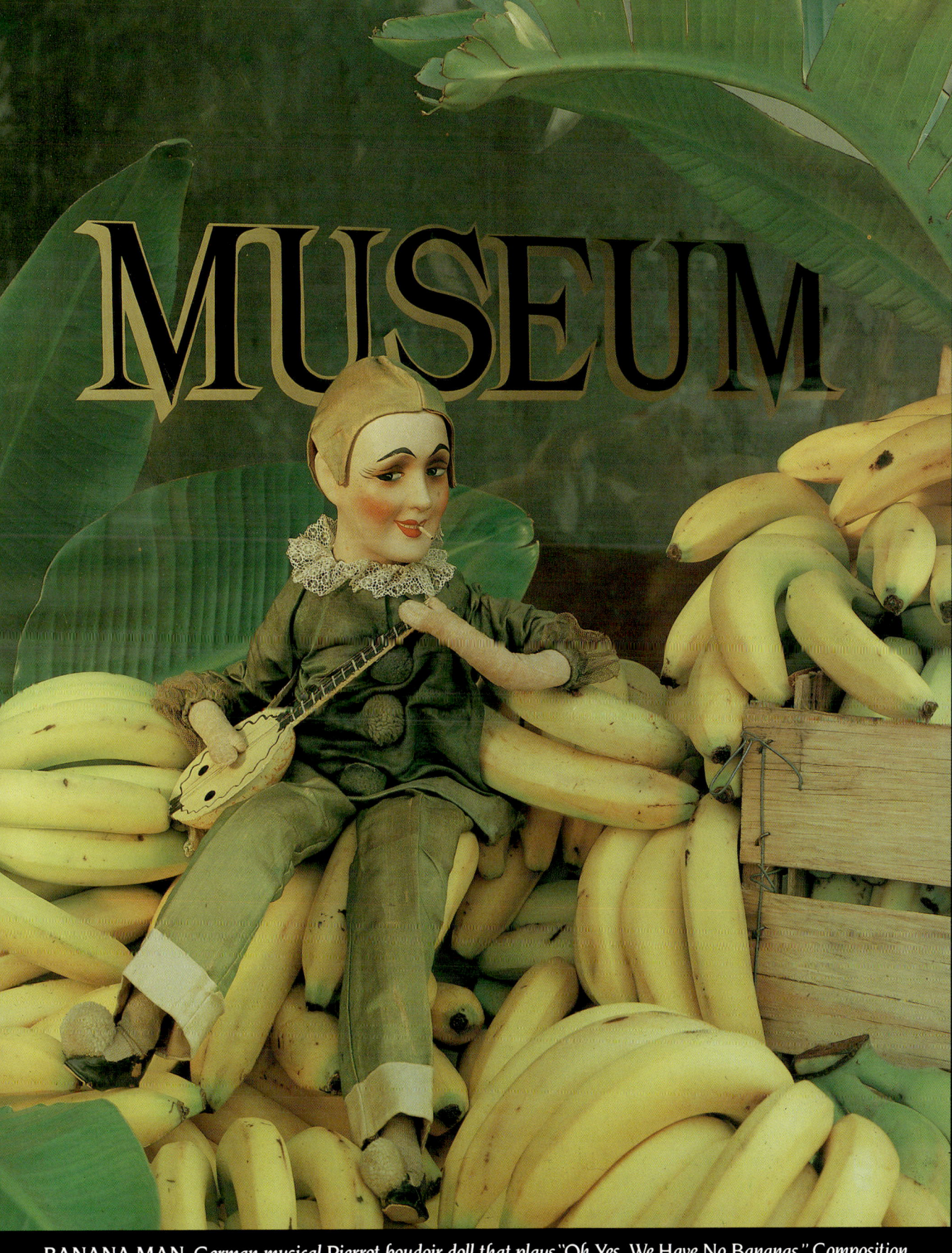

BANANA MAN. *German musical Pierrot boudoir doll that plays "Oh Yes, We Have No Bananas." Composition face and all cloth body, stands 25" high. Farago Collection.*

Romance is a game for fools,
I used to say;
A game I thought I'd never play.
Romance is a game for fools,
I said and grinned; then you passed by,
And here I am throwing caution to the wind.

Fools rush in
Where wise men never go
But wise men never fall in love
So how are they to know.
　　—LYRICS BY JOHNNY MERCER, "Fools Rush In"

© 1940 (Renewed) WB MUSIC CORP.
All Rights Reserved. Used by Permission.

HARLEQUIN HONEYMOONERS. Two American Blossom harlequins with silk hand-painted faces and elaborately patterned costumes, 30" high. One was found at an auction and the other in a private collection. Erté's comment when he saw them was, "They make a perfect marriage". Farago Collection

> He who hath not a dram of folly in his mixture hath
> pounds of much worse matter in his composition.
>
> —CHARLES LAMB

CIRCUS MAGRITTE *features a German wax-faced clown (center), which was probably the prototype for the later composition-faced American Anita dolls to left and right. They all have cloth bodies. Pierrot figure, on left, has pajama bag culottes with a club, heart, spade and diamond on them. They stand 31" high. Farago Collection.*

How happy is he born and taught,
That serveth not another's will;
Whose armour is his honest thought,
And simple truth his utmost skill
—SIR HENRY WOTTON

HURÉT JESTER. A French Hurét found in very poor condition and redressed as Queen Elizabeth was refurbished by doll artist Mariel Marlar. She gave the doll a new persona of "jester" when she redressed the doll, created a new sheepskin wig and a matching marotte for him to talk to. He stands 18" high, with bisque face, pewter hands and fully articulated wooden body with a swivel waist; he sits beneath a very rare Tiffany lamp. Richard Wright Collection.

*The man who lives by himself and for himself
Is likely to be corrupted by the company he keeps.*
—CHARLES H. PARKHURST

SOLEMNITY. A *petulant Pierrot with a rare Lenci face. He is all felt and is 26" high.* Richard Wright Collection.

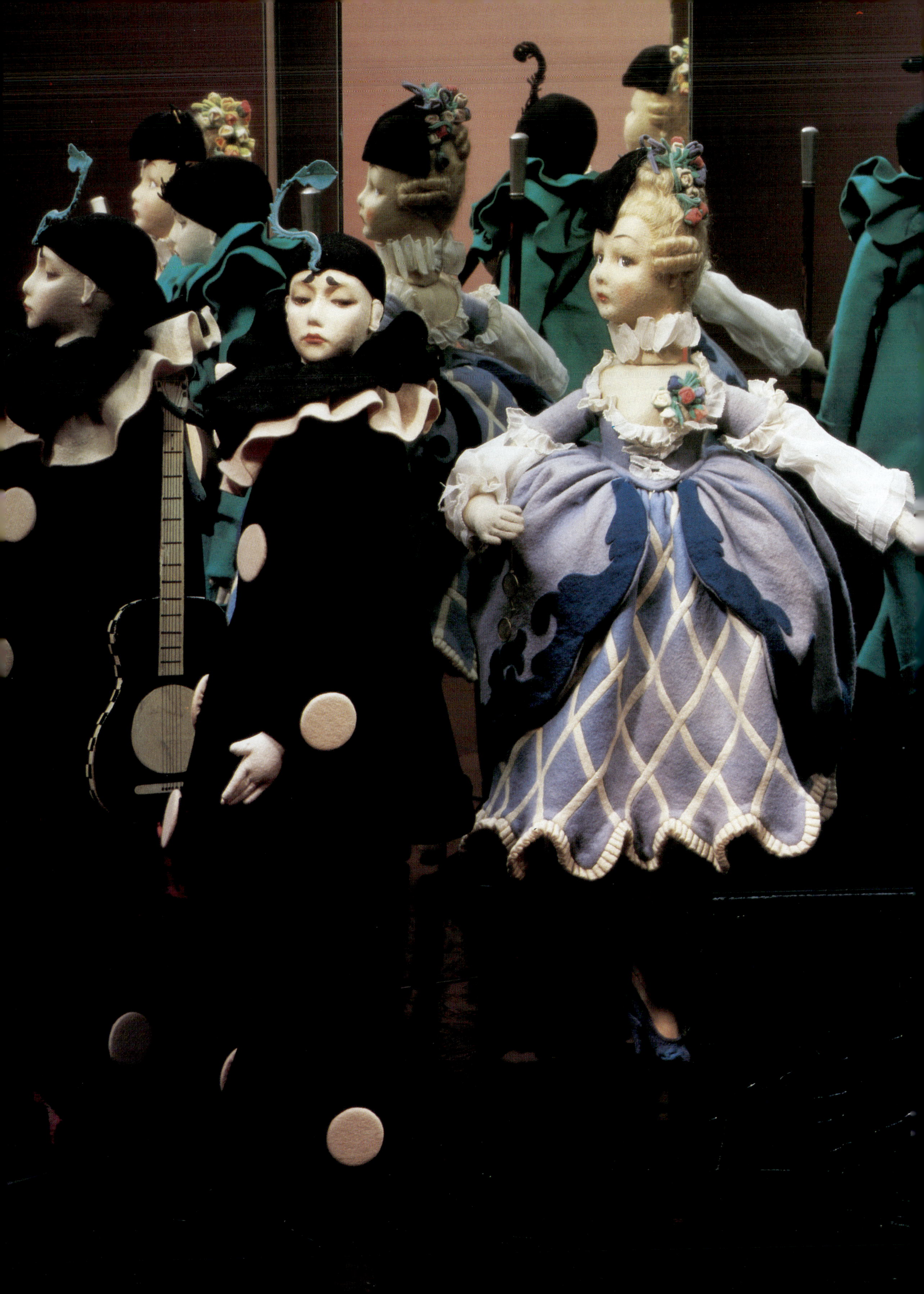

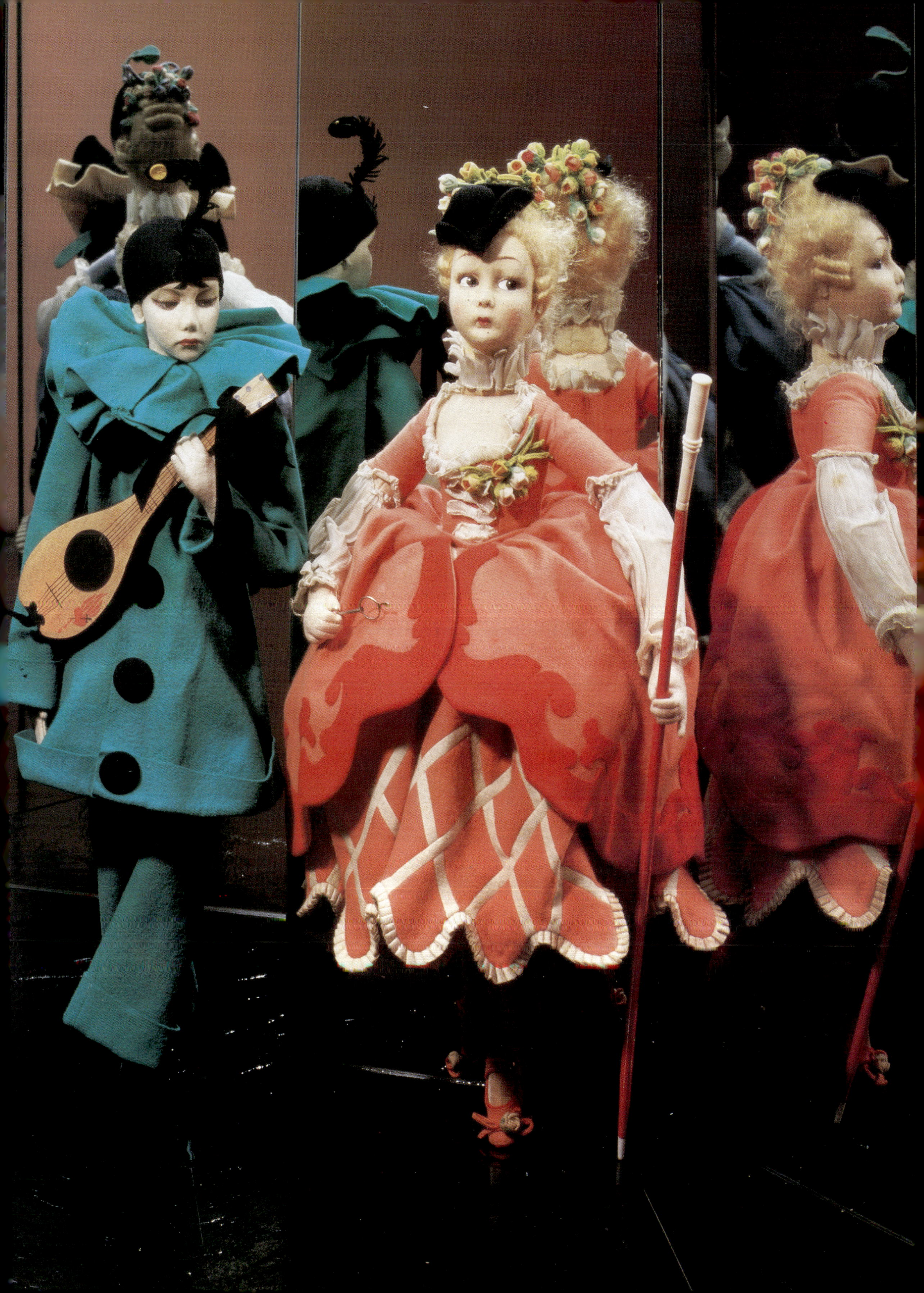

After dinner we made a pillow of my shoulder, I read to him and my Beloved slept.

—DOROTHY WORDSWORTH

LE BAL (*previous page*). Lenci Pierrots and Columbines celebrating in a grand ballroom. They are all felt and stand 25" high. Farago Collection.

Don't part with your illusions. When they are gone, you may still exist, but you have ceased to live.

—MARK TWAIN

LOVERS, PIERROT AND PIERRETTE. *These two tattered souls have such a sensitive patina that one can only love them more in their disheveled state. French flocked-faced, dressed in white satin with large flat black buttons, they are 35" high. Farago Collection.*

"Imagination is not a talent of some men but it's the health of every man."
—EMERSON

CLASSIC CLOWN. All fabric, made in Florence, Italy in the 1950's, labeled Isotie Artistic Dolls. Gladyse Hilsdorf Collection, Fayetteville, N.Y.

If you believe your own claim to miracle doing and are sincere in your work, you are bound to succeed.

—HOUDINI

HARLEQUIN FORTUNE TELLER. *An all felt Lenci Harlequin is original except for his hat, 29" high. Farago Collection.*

The king who fights his people fights himself.
—TENNYSON

SPIDERMAN. A rare English Norah Wellings Jester mischieviously winks as he traverses his web. He wears black velvet with red puffs and trim, 22" high. Farago Collection.

Snobbery is but a point in time. Let us have patience with our inferiors. They are ourselves of yesterday.
—ISSAC GOLDBERG, "Tin Pan Alley"

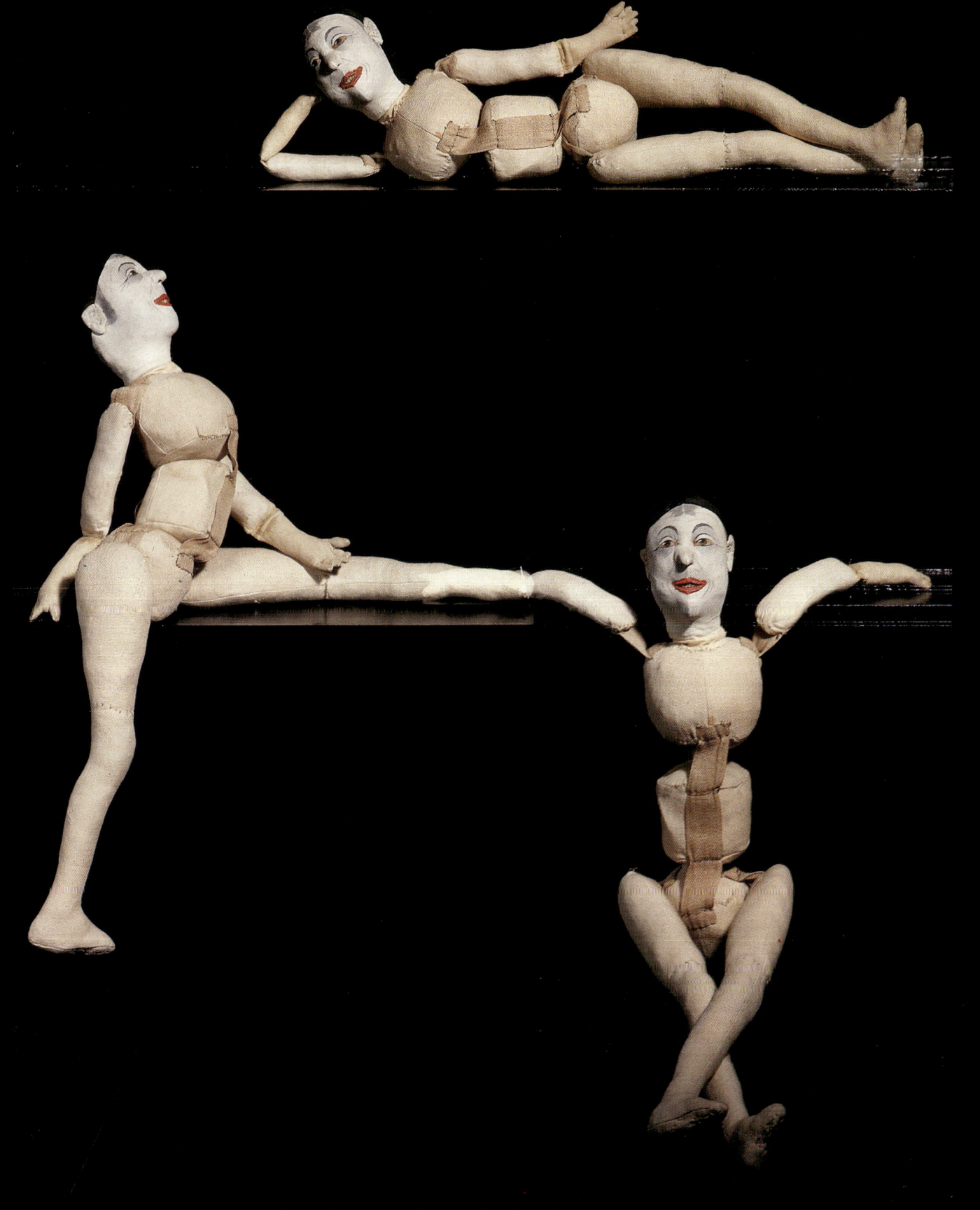

MUYBRIDGE PIERROT. Shows three views of a very flexible doll. Made entirely of canvas and sports a unique construction. He may be German and the only one of its kind, since the hand-painted face and construction reveal that he is definitely not manufactured. Farago Collection.

Sensual pleasures are like soap bubbles, sparkling,
 effervescent.
The pleasures of the intellect are calm, beautiful, sublime.
Ever enduring and climbing upward to the borders of
 the unseen world...

—JOHN H. AUGHEY

JUGGLING JESTER. An all cloth European jester, stuffed with excelsior, has a blue felt heart sewed to his silk and corduroy outfit, 29" high. Farago Collection.

First we make our habits then our habits make us.

V. Other Adult Amusements

THE NARGILEH SMOKER *introduces a number of adult amusements and other oddities.*

Entrance to the Musée National, Monaco, the principality where Prince Rainier reigns. The Prince Rainier and Princess Grace dolls are on page 80. The museum houses one of the finest doll and automata collections in the world.

The luxury of amusement began to flourish when survival and self-defense became less crucial. Warriors, after the Crusades, brought back furniture, novelties, and art objects from their conquests. Decorating the castle with these new victory items became a court event. There was also leisure time to have visitors come to celebrate and admire these new acquisitions.

Ceremonies honoring the victorious, where warriors would be knighted, gave women a chance to participate as they never could have during war time. Women were thought to be slightly better than useless since they could not fight. Now they had a function: to dress up and attend the ceremonies. Having a natural bent toward social grace, the ladies enhanced the events and found a desirable purpose. Helping to decorate and host the festivities became a castle preoccupation when the excitement of war died down.

Romance and chivalry became the medieval theme as the women did everything possible to dazzle the heroic knights. The women were also given the job of appointing troubadours to the court. The troubadours were poet-song writers who lived in the court and were summoned to entertain at every opportunity. They would write about courageous men and deeds of battle as well as lyrics of love. The heroes the

Love knows the art of adorning itself with artificial charms for the great game of love. Illustration by George Barbier

Art Dolls

This is a cloth head of a French doll marked "Rosalinde" on the foot, 30" high. Farago Collection.

THE MASK

The mask, as an accessory, was used to heighten romantic intrigue. Louis XVI was known to wear a mask during a whole evening and dance with whomever he pleased without ever divulging his true identity.

THE LANGUAGE OF THE FAN
(found in a Bourbon Street perfumery)

Handle to lips = Kiss me.
Fan in right hand in front of face = Follow me.
Fan in left hand = Desirous of acquaintance
Twirling in left hand = Get lost!
Drawing fan across forehead =
              We are being watched.
Pulling fan through hand = I hate you.
Fanning slowly = I am married.
Twirling fan in right hand = I love another.
Fanning fast = I am engaged.
Closing fan = I wish to speak to you.
Drawing fan across eyes = I'm sorry.
Letting fan rest on right cheek = Yes
Letting fan rest on left cheek = No
Open and shut = You are cruel.
Open wide = Wait for me.

Pierrot fan advertising French champagne. Farago Collection.

songs were written about were totally unskilled in social graces and often illiterate. Eventually, the songs of the troubadours inspired the once-crusty warriors to become poetic, to write their own lyrics and melodies, and to woo the ladies in this new, gracious atmosphere.

The competitive spirit of war was channeled into the conquest of winning a lady's affection. Gifts were used as items of persuasion. Delightful accessories were given and used to provoke romantic encounters. The Commedia dell' Arte, with its theatrical costuming and masks, made masquerades a favorite in the courts. The passionate party-goers often donned masks and fans to heighten the mystery of their interludes.

182

Art Dolls

Hermaphroditic Court Doll attributed to the Louis XVI period, carved of wood, 19" high. Gladyse Hilsdorf Collection, Fayetteville, N.Y. "He who, being a man, remains a woman, will become a universal channel. —Lao-Tzu

One of the most provocative dolls in all doll history is probably the Court doll. It is an erotic wooden figure with moveable joints, sexually explicit details sometimes anatomically correct and sometimes overly endowed with one or two sets of male genitalia as well as female breasts. It's difficult to know what was actually done with the dolls, but one can imagine. They were perhaps the first sexual kinetic playthings. Court dolls have been attributed to the 1500's, although some speculate that the joints are too modern to have been created during that period and were made during the 1900's. Since a perfectly working battery was found among artifacts discovered in ancient Egypt, it is certainly possible that court doll construction could have been managed as early as 1500.

By the middle of the 1800's in France, some very talented men were making toys that moved with only a turn of a key. Deschamps, Vichy and Lambert were the *automata* artists who delighted the extremely rich. Even holograms and lasers cannot parallel the excitement these moving sculptures provoke. Some of them actually breathe as they change expressions from moment to moment.

Mainly, the accessories shown in this chapter are from the mid-1800's through the Art Deco period... They are the amusing accouterments of courtship and entertainment for adults who refuse to stop playing and who probably will never grow old.

Automata

During the mid 1800's, Parisian artisans were intensely involved in the making of mechanical dolls. The most famous sculptors of automata were Vichy, Lambert and Deschamps. The Musée National in Monaco has many of the rarest automata in the world today.

THE PAINTER-POET. (above) Only one of its kind, is a priceless sculpture by Vichy, 1875, 34" high. His face and hands are made of composition, and his eyelids are made of soft leather, enabling them to flutter. He draws in his book, admires his work, shows it to his audience and sighs with self-satisfaction. Musée National, Collection de Galéa. Photo: Stephanie Farago.

EQUILIBRIST CLOWN (left) by Vichy, 1875, 30" high. Musée National, Collection de Galéa. Photo: Stephanie Farago.

THE NARGILEH SMOKER *by Lambert, 1885, 17¾". Plays Mozart's March of the Turks while he sips coffee and smokes his hookah. Musée National, Collection de Galea.*

PIANO HARPIST by Vichy (circa 1870), 30" high. She is like a vision in sherbet, as she rhythmically flutters her eyelashes to the tune she gracefully plays. Musée National Monaco, Collection de Galea.

> Nothing is less in our power than the heart, and far from commanding we are forced to obey it.
>
> —JEAN JACQUES ROUSSEAU

PIERROT WRITER by Lambert, 23" high. Musée National, Collection de Galéa. He writes a letter dated Paris 26 May 1874 ... "My Dear Columbine, The time is going so quickly. Even though the years have no meaning, I love you just as much as I always have. Your Pierrot from yesterday and today."

Art Dolls

At the end of the 1800's the first amusement houses were built to entertain a paying public. The Musée Grevin, Paris' great wax museum, was built to display some of the best surviving mannequins from that time in conjunction with a house of mirrors, which is a limitless and unforgettable vision.

Many fine museums still house these man-made wonders and novelties. It is our pleasure to share with you some of the most memorable.

Musée Grévin's Palace of Mirrors was the masterpiece of the 1900 Paris exposition. Built by architect Emile Henard, the mirrors revolve to reveal three scenes: the Temple of Brahma, Enchanted Forest and Feast in l'Alhambra. The room has forty-five lighting possibilities.

Entrance to the Musée Grévin, one of the most elaborate wax museums in existence. Like a seventh wonder, it is a "must see" to any tourist visiting Paris.

A wax, life-size Asian woman in Musée Grévin's Palace of Mirrors.

A wax, life-size black woman in Musée Grévin's Palace of Mirrors.

Musée Grévin's foyer's four corners are graced by these life-size, wax figures from the Commedia dell'Arte.

Columbine

Harlequin

Pierrot

Polichinelle, a wax, life-size mannequin. (Detail of one of the four corners of the Musée Grévin.)

Doll Muff

Accessories

Throughout the Art Deco period, the doll image was touted far beyond the confines of statuary, amusement, or adornments for the boudoir. Stylists transformed everyday objects such as hat stands, handbags, cigarette dispensers and lamps into facsimiles of dolls—dolls with function as well as fashion. At no time in history has the doll image been more celebrated.

Erté, an artist of great distinction during that time, illustrated a variety of imaginative ways to incorporate dolls into accessories, as shown here. The following pages reveal more Deco doll sculptures and accessories that appeared until 1940.

Doll Hat stands

APR. 1919

Doll Purse

Doll Purse

Doll Purse

To carry with her evening costume and to provide a carry-all for her little nothings, Erté makes Madame a bag of silver tissue and semi-precious stones. The bouffant bloomers of tissue form the bag.

The matron who would be correct may carry a doll dressed exactly like herself. The skirt of mauve and white taffeta embroidered in orange is an efficient hand-bag. Even the hat is not forgotten.

Frills of black and white taffeta form the doll's petticoat and in turn a bag for the debutante. The doll's face is hidden by her bonnet. The ballon is really a jade ball attached to a thin chain.

Doll Purse

THE FRAGILE BISQUE DOLL FINDS A NEW ROLE
—Harper's Bazaar, 1921

This page and opposite page courtesy of Harper's Bazaar/Hearst Corporation © Sevenarts Ltd., London.

Accessories (above): feather duster 2½" high, doll-headed cigarette dispenser, match-box with small candle holder, doll-headed clothes brush, 3½" high. Farago Collection.

Doll perfume bottles (below), 10½" high (center front and back). Doll heads separate from bodies, and necks serve as perfume stoppers. Other dolls (seated left and right) are mascots used to adorn cars or given as party favors. Farago Collection.

Doll head wig and hat stands. A German hat stand, (left) 11″ high, American Anita Pierrot (left of center) 11″ high, French hat stand (center left) 13½″ high, and one marked Japan (far right) 11″ high. Farago Collection.

German bisque figures (above) pose on top of a glowing lamp, 6" high. Farago Collection.

Deco boudoir sculptures, (below). Center one stands 8" high and is a lamp base while one on far right stands 7½" high and strongly resembles Dietrich. Figure on left, 6½" high, is all cast metal on marble base. Other two are cast metal with carved ivory heads and hands. Farago Collection.

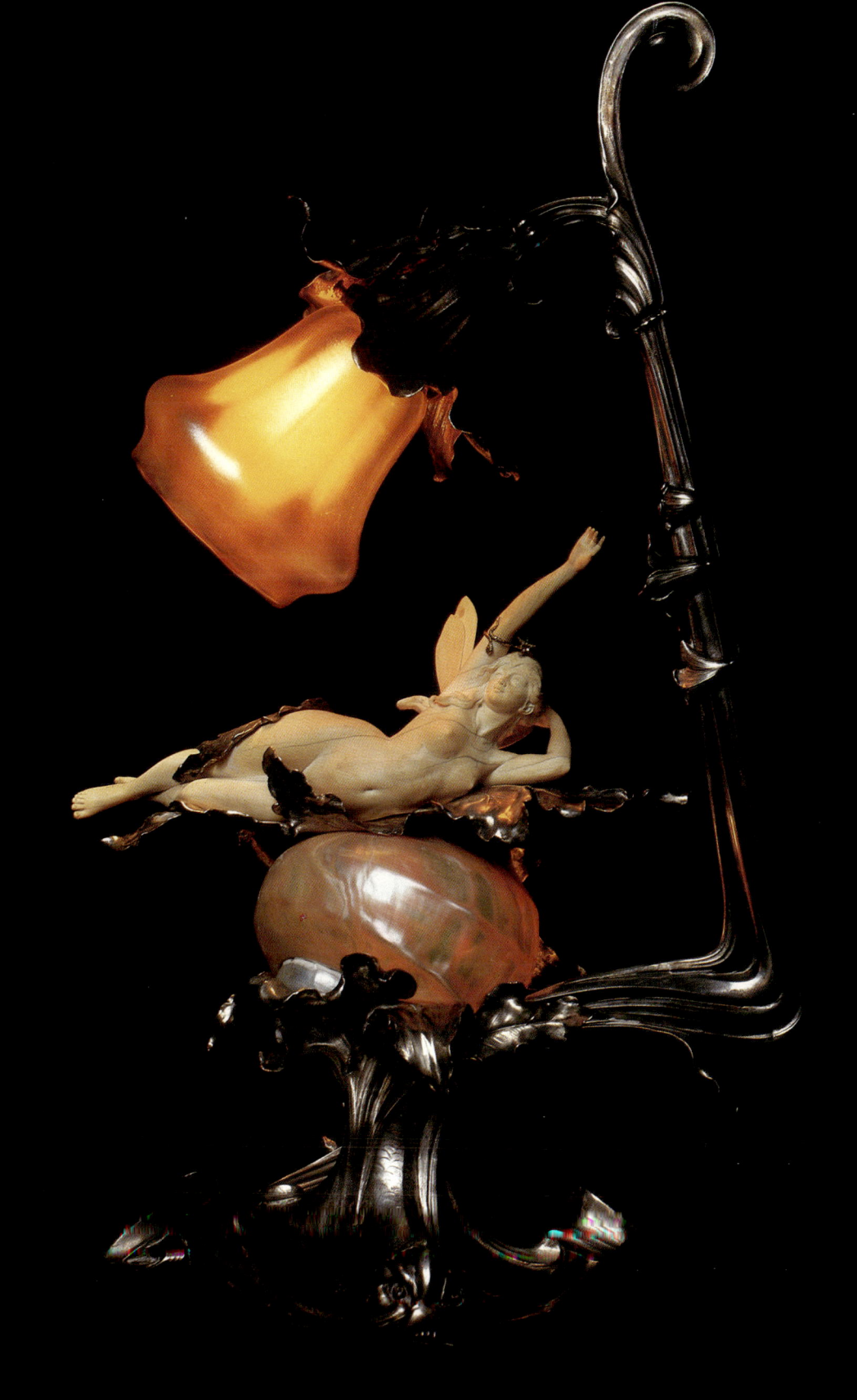

A Tiffany lamp with dramatically posed, 12" ivory nude. Richard Wright Collection.

French paper dolls (above) 1840 – 50. Barbara Whitton Jendrick Collection.

Boudoir dolls (left) were often-times constructed from mask faces such as these. Farago Collection.

Eve (left) holding the apple,
by Dressel Kister and Co.
(Passau Bavaria, Germany)
1900 – 1930, 6" high.
Strong Museum, Rochester, N.Y.

Lady with fan (right), German pincushion 8⅝"
high, 1900's. Strong Museum, Rochester, N.Y.

Pierrot accessories include four Pierrot soap dishes or ring holders (center). Pierrot head with threads on neck is a bottle top (front center). Open mouth is a tape measure (middle center). Pierrot candy box 6" high from seat to head (back row center). Pierrot and dancer powder puff (back left), Pierrot lamp base (left of back center), Pierrot powder puff handle (back right), also two Pierrot heads, one wax (right of back center) and the other two-piece porcelain.

Doll-headed hangers (above) were used in the boudoir. Farago Collection.

Feathered lingerie bag, (right) Farago Collection.

German pincushions and nudies. Figures (front left) 5" high are pincushions (notice holes on waist to sew onto cushions). All dolls from center to right have human hair wigs; back row shows ornately finished pincushion dolls. Farago Collection.

Candy boxes with portrait of Charlie Chaplin, 7½" high, and flapper girl, 8" high. Farago Collection.

Various Deco jars. Pink alabaster with painting (front center), Hertwig matching pink porcelain jar heads (left), German lidded box with bathing beauty on top, (right). Farago Collection.

German bisque ladies, 1900's. Figure (front center) is 10" high, figure (left back) 14" high and has "Pandore" inscribed on metal strip around base. Figure (left of center) 11" and seated figure 9½" high. Farago Collection.

Art Dolls

FINE ARTIST MAKES LASTING STATEMENT WITH DOLL IMAGERY

HANS BELLMER

Of all the myriad uses of the doll, the fine artist Hans Bellmer stands out as one who used the doll symbol brilliantly to express his horror against the rise of Nazism in Germany in the 1930s. It was his mission to reach into the fantasy world of the adolescent child's realm to rejuvenate himself and his art while making "an artificial girl", a rebellious statement against the demonic power structure that was beginning to pervade his very private world.

Bellmer's attraction to the doll began upon his seeing Offenbach's opera of *Tales of Hoffman* in 1933, a Pygmalion/Svengali type of story wherein Hoffman finds himself in love with the mechanical doll, Olympia. Bellmer ceased whatever activity that might have been construed as useful to the State and began his construction of the artificial girl.

During 1933, he made friends with the famous dollmaker, Lotte Pritzel, who, along with her husband, encouraged his interest.

1934 was the year he connected with the Paris Surrealists and met André Breton, Paul Eluard, Marcel Duchamp, Max Ernst, Yves Tanguy and others. He became so enraptured with his doll work, he wrote and published *Die Puppe*, a book on his work with the doll image. In 1938, he left Berlin to settle in Paris.

The artificial girls are perhaps the most erotic sculptural statements of this century. More importantly, they were expressions of Bellmer's "inner life." His works are surrealistic with intriguing psychological overtones. He probed the existence of an unexplored physical world which he called "the physical unconscious."

Bellmer said "...the Doll, who despite her easy-going and infinite docility, surrounds herself with a heartbreaking reserve."

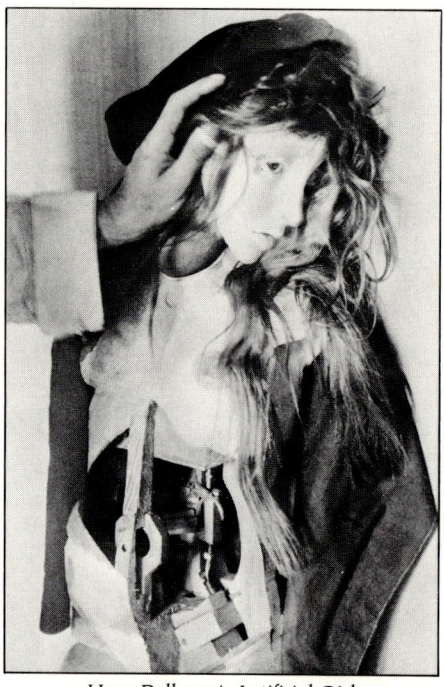

Hans Bellmer's Artificial Girl.

VI. Encore

Brigette Starzcweski Deval's wax-over-plaster wood nymph seems imbued with an inner light.

DOLLMAKING TODAY... A CONTINUING ART FORM

Throughout history and within all the cultural arts, there has always been a prodigious place for the doll. Currently, doll collecting rates second among hobbyists in the world. The demand and interest today is so great that many new dollmakers are beginning to emerge once again. Some of them are demonstrating a fresh, innovative creativity, while some are content to copy the past or to cast it over and over again.

Perhaps the single thing that made dollmaking a mediocre commercial enterprise, instead of a fine art, was the factory assembly line. While providing jobs to the unskilled, it also diminished the need for talented artisans to make one-of-a-kind, expensive artworks and gifts by hand.

The artists represented on the following pages are current dollmakers. They span the globe from the United States to France, Germany and Japan. Although the representation of modern day dollmakers is limited in this edition to just a few, there are probably enough brilliantly talented dollmakers to comprise another entire book. Each of the artists presented has his/her own materials and way of working, and each has created images reflecting humor, sensuality, antiquity and grace. We would consider this book a great success if it inspired fine artists to see the sculptural art form of dollmaking as an opportune, creative field.

BRIGETTE STARZCWESKI DEVAL
(Doll *sculpture pictured on previous page*)

My introduction to this talented young woman took place in Monaco, at the World Doll Congress in March 1984, where she was exhibiting and selling her dolls. Like a one-woman band, Brigette Deval seemed to orchestrate the many demands placed upon her, as she nursed her newborn baby, spoke a number of different languages and attended to the inquiries of many interested customers.

I was enchanted by the lovely dolls of children, made of wax over plaster, but what I found to be the most captivating was a lone female figure, nude from the waist up with skin so transparent it looked like human flesh. I asked her why this was the only "adult" image in her display and she said that this particular doll was a gift she was making for her husband.

Brigette Deval first developed an interest in dolls as a child when her father took her to visit the doll museum in Munich. Her imagination was irrevocably captured when she saw the marionettes, moving as they seemed to magically come to life. The inner light of the wax angels, with their tiny features, especially delighted the eye of the impressionable little Brigette.

Her first doll attempts were hampelmann, little men with pull-strings that automated the legs and arms. She often made them for her friends and family as gifts while she was growing up. But she was not impressed with her first attempts and for a number of years attended college with no thought of dollmaking. Once again, she gravitated toward the doll world when she began to read about dolls while at school. Finally she began to sculpt again. Now, with the skill of an adult, she works to incorporate her original child-like vision of that inner light coming from the dolls. Her dolls can be seen at the Munich Shop in the Trump Towers in New York.

Nicholas Bramble at work on a life-size male mannequin.

NICHOLAS BRAMBLE

Nicholas Bramble, a native of Norfolk, England, says of his first visit to London at the age of eight, "The lights went on." After visiting Madame Tussaud's famous wax museum and seeing a movie of Ulanova in the Bolshoi Ballet's "Giselle," he began a rhapsodic affair with ballet and dollmaking.

From childhood to adulthood he stud-

ART DOLLS

Gibson Girl, patterned after the illustrations of Charles Dana Gibson, is hand-painted bisque with mohair wig and stands 27" high.

...ied both ballet and art. His natural sense of grace became an integral part of his art-works. As a student and dancer, he felt compelled to make what he could not afford to buy.

He was invited by Bernard Tussaud, the last survivor of the family, to work as an apprentice in the wax museum. At the same time, he became a wig-maker for the theater and danced with the English National Opera.

He created his first wax doll at age

A luscious mannequin head by Bramble.

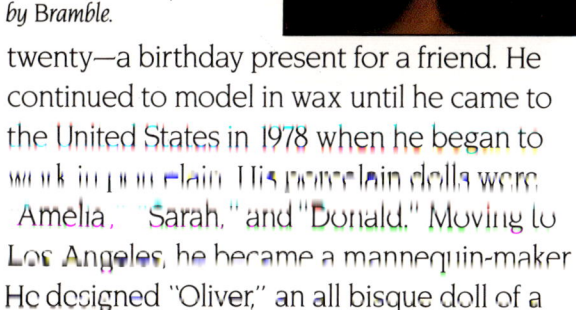

twenty—a birthday present for a friend. He continued to model in wax until he came to the United States in 1978 when he began to work in porcelain. His porcelain dolls were "Amelia," "Sarah," and "Donald." Moving to Los Angeles, he became a mannequin-maker. He designed "Oliver," an all bisque doll of a lovely little boy, for production by Janice Cuthbert. "Lily" an elegant, turn of the century lady, was designed with the drawings of Charles Dana Gibson in mind. Being a rather multi-faceted individual, with mannequin-making and wig-making as his constant sidelines, his dolls are wonderfully beautiful and rare.

CHARLOTTE V. BYRD

Charlotte Byrd is one of those people who sees something she cannot possibly afford and goes home and makes it from scratch. Her mother Rose, a dedicated doll collector, asked Charlotte to research some of her dolls at the doll museum. When she came upon the works of the famous Martha Thompson, (creator of the Prince Rainier and Princess Grace dolls pictured on page 80) her love affair with dolls officially began. Realizing these dolls were far beyond her means, she went right home, created a porcelain doll and baked it in a kiln, only to find her first progeny had blown to bits. Undaunted, she continued in her efforts, and now, nine years later, Byrd has an extensive line of dolls she has created.

Although she is a self-taught doll artist, she received her degree in art from Auburn University and spent a number of years as a graphic artist and technical illustrator. With her first 100 pounds of plaster she experimented with mold making until she finally started getting results. Next she began learning to paint china, make wigs, as well as design and sew costumes. Her hand-sculpted dolls include celebrity portraits of Albert Einstein, Abraham Lincoln, Helen Keller, clowns, children and the magnificent Merlin the Magi-

ART DOLLS

Charlotte V. Byrd's bisque "Merlin the Magician" is a sophisticated sorcerer.

cian. Each doll takes her about 150 hours to complete. Sculpting in plasticene clay, polyform clay or ceramic clay, she then makes a plaster mold and pours the porcelain.

The Atlanta-born belle began her career in May of 1978 and by October of that same year, she was voted into the Original Doll Artists Council of America (ODACA). In Christmas of '83, her dolls were on exhibit at the Huntsville Museum of Art. Although it probably would have been less expensive to buy one of Martha Thompson's dolls, no one can question the commercial success of Charlotte Byrd's creations. She is certainly a woman who realizes her visions.

MARCEL CHARLES GAICHET

While on route to visit Erté, we were fortunate to visit the Tuturlu doll shop of Marcel Charles Gaichet, tucked away behind the main streets in Montmartre. We spent a day in this unique shop filled with Marcel Charles' nostalgic works. His dolls are truly boudoir dolls, the natural offspring of their French ancestors.

At an early age, Gaichet fell in love with a salon doll belonging to his uncle. When his uncle died, he asked if he could have the doll, but the family refused him. This unrequited

Celebrity dolls and fantasy figures are authentic, contemporary boudoir dolls created by Marcel Charles Gaichet. They can be found in his unique Parisian doll shop, "Tuturlu."

love brought about an inner rebellion that caused Gaichet to make dolls the focal point of his entire life.

First, he became a china painter and sculptor in a little factory in Provence, outside Paris. In a Perpinean, town he studied the tradition of sculpting Crèche figures at the Ecole des Beaux Arts.

Each boudoir doll employs an identical, plain white bisque face, which is individually hand-painted to distinguish it. The hands, forearms, lower legs and feet are all made of the same white bisque used for the heads. The bodies are soft sculptures which are wonderfully graceful since they have jointed elbows, waists and knees. Besides his penchant for sculpting and china painting, he is also a fine designer of fashion and sews his original designs for all the dolls he crafts.

His characters are usually romanticized stars and theatrical personalities such as Marlene Dietrich, Charlie Chaplin, Brigitte Bardot, Marilyn Monroe, Edith Piaf, Sarah Bernhardt, Marcel Marceau, as well as some fantasy creatures which include fairies and witches. If Paris isn't your next stop, Marcel Charles Gaichet's dolls can be found in Los Angeles at the PePé Boutique.

TITA VARNER

Tita Varner represents the modern American, creative homemaker, whose artistic talents enhance everything she does. Her many skills include oil painting, flower arranging, clothing design, ceramics and pottery. Along with the help of her husband, Tom, who is mold maker, bookkeeper and photographer for her line of TiTa-Dolls, she has established a cozy cottage studio that is

Tita Varner's "Columbine" is an elegant representation of the sweetheart of the Commedia dell' Arte.

lucrative, creative and fulfilling. Varner says, "There's just something about pouring all my energy into a new design, and when I finally hold the completed doll in my hands, it's like magic!"

Her award-winning dolls are all kiln-fired at least five times, at the highest possible temperature to achieve a soft, natural sheen only found in fine porcelain. The limited edition TiTa-Dolls range from 10" to 24" and the eyes are china painted to give them a life-like expression. "Columbine" is produced in four sizes from 14" to 24" and is also available as a bust.

Art Dolls

PETER WOLF

Growing up in a fairy tale city is as good a reason as any for a young man to become a maker of dolls and fantasies. Born in the ancient town of Wurzburg, Germany, full of medieval surroundings, castles, basilicas, turrets and classical legends, Wolf began experimenting with his creative talents while still in high school. He made costumes and props for theatrical productions, worked in fashion design and created ornaments for Christmas to sell in giftshops.

It was during Christmastime that he met with the flirtatious glance of a Harlequin doll in an antique store in Munich. This new love sparked his desire to make his own version of a Harlequin. When his first creation sold immediately, his future was certainly an adventure beginning to unfold.

Each doll is sculpted by hand from a sculpting material. The heads, arms and legs are molded, fired and painted. A padded wire armature underlies the construction of each doll. The clothes are all exquisitely designed, tailored and decorated by the nimble fingers of Peter Wolf. His figures include characters and creatures from medieval times, the Roaring Twenties, the theatre and rock and roll. Elegant ladies, cats and notables as unique as Boy George are among his repertoire.

Peter Wolf's Art Deco lady and catman walking their pet mouse. Adorned with natural furs and flawless tailoring; they stand 15¾" high.

A Pierre Imans mannequin attributed to Erté.

ERTÉ

The Erté Doll was done in collaboration with Shader's China Doll company. She is regarded to be a fine art doll because she is a work of art and was designed and signed by one of the world's greatest living artists, Erté. Famous for his Deco period styling of lavish costumes, sets, lithographs, and serigraphs, Erté has enjoyed a revival of his work in the last twenty years.

When Erté was ninety-one, he designed the Erté Doll as a limited edition of 300, including thirty numbered artist's proofs. Made of fine porcelain, all molds are destroyed upon the completion of the edition. The costume is created of richly extravagant fur and velvet and is an exquisite realization of the craftsmanship attributed to the works of Erté and Shader.

HIROSHI HORI

Hiroshi Hori is a direct descendent of a dollmaking mother, a ventriloquist father (who made his own glove-puppets to entertain the neighborhood children), and a grandmother who was a traditional Japanese dance teacher. In Japan, the cultural emphasis of the doll touches everyone's life. Whenever a child is born, male or female, a doll is given as the first birthday present. In fact, Japanese dolls have served in ancient tomb rituals, exorcisms, memorials to children who have died and yearly festivals known as "dolls day." Before modern times, a woman's only personal possession might have been the doll she received at the time of her birth.

In Japan, it is no wonder that a man would own or create dolls for a living. It is no surprise either, that Hiroshi Hori, after getting a solid education in economics, could no longer resist the calling to be a genuine *ningyoshi* (a master dollmaker or doll-manipulator.) At twenty-seven, he has brought all his talents together to become one of the most eminent modern men to continue the 18th Century tradition of *ningyo jyourui*, "the ballad with dolls", which later became known as the Bunraku. The Bunraku puppets were so large it took three men to manipulate them.

Truly a trendsetter, he not only fulfills the functions of three men doing the manipulation, he is also a master dollmaker, kimono designer (dyeing and sewing), a stunning performer, choreographer and lighting director. The dolls are worked by Hori with two rods: one which moves the hands (in Hori's right hand), and the other moves the head (in his left). Together, the doll and an invisible Hori dance gracefully through the staged lights illuminating the expression of the once-lifeless doll figure.

At age twenty-seven, Hiroshi Hori is one of the most eminent modern men to continue the 18th Century tradition of ningyo jyourui, the ballad with dolls. His life-size dolls, weighing from 22 to 44 lbs., are multi-media soft sculptures with cloth faces; hands are made of harder material.

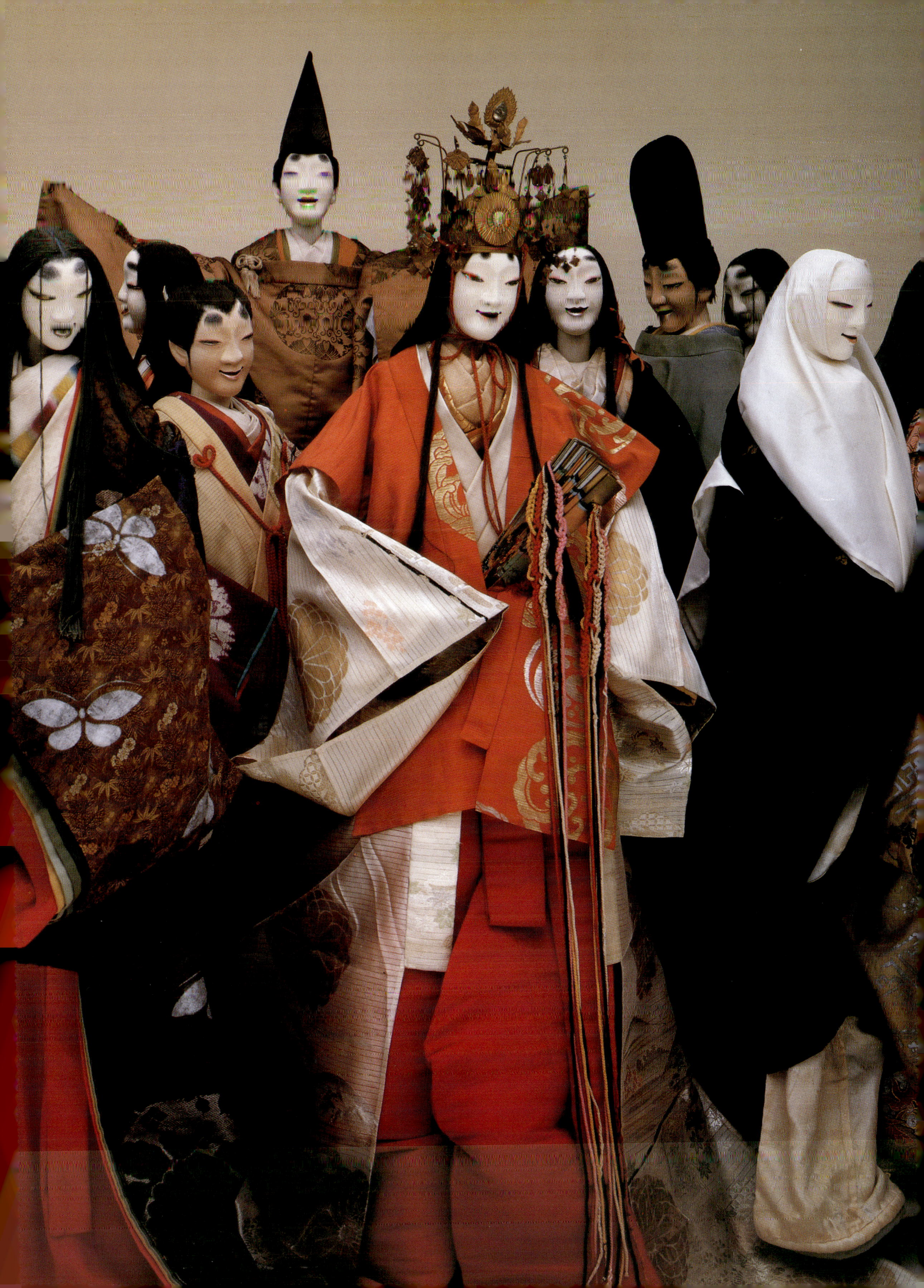

Art Dolls

SARA GOLDSMITH

Sara Goldsmith was reminded constantly by her father to do "the sensible thing." While her brothers struggled in the liberal arts, she became an accountant. Burying her true self in a work-a-day world, she eventually began to work with her hands, assuaging her guilt by calling it "a hobby."

Beginning in the 1930's, Goldsmith enrolled in an arts and crafts class at the Newark Museum. Her timidity soon vanished when her instructor praised her very first mask of Topsy. Next, her sculptural portrait of Abraham Lincoln met with even more praise, which was surprising only to Goldsmith herself. Her instructor suggested she create a Nativity scene based on a painting of Correggio. It was displayed behind a gold frame as a three-dimensional rendition of the original oil painting.

Her one-of-a-kind portrait dolls became her forte; the likenesses she created were uncanny. In 1947 the Museum of Science and Industry in New York City exhibited Goldsmith's rendition of Princess Elizabeth dressed in authentic, regal wedding attire.

Detailed research is a pre-requisite for Goldsmith's portrait dolls. In the case of the "Princess Elizabeth" doll, she studied drawings of her done in England. The "John Noble" doll wears a clerical outfit, which is an exact copy of one he actually owns. Most of her dolls are one-of-a-kind, and range from 9" to 20". The heads and torsos are made from platilina; the molds are made of plaster of paris and cast in plastic wood. The hands, feet and legs are made from carved wood or made of wire with cloth or felt.

A dollmaker for over fifty years and member of ODACA, she found that the "sensible thing" for her has been to make dolls. And, breathing life into her dolls is certainly her favorite vocation.

Goldsmith's "John Noble" doll is a beautiful and stunning hand-painted likeness of the curator of the Doll and Toy Department of the Museum of the City of New York.

VAN CRAIG

An outrageous and hilarious art form emerges from the Manhattan studio of sculptor Van Craig. A theatrical performer, as well as a set and costume designer, Craig creates soft, flexible sculptures of papier maché. He uses vintage clothing, feathers, beads, rhinestones, imported German glass eyes, antique teeth and human hair to materialize his menagerie of merriment.

His "Dramatis Personae", as he likes to call them, are one-of-a-kind sculptures ranging from 18" to 3½' and bring prices as large

Art Dolls

Van Craig's God Wotan (above), a character featured in Wagnerian operas. He fraternizes (below) with his assorted characters and creatures.

Craig's bizarre brood includes Judy Garland as Oz's Dorothy and Bert Lahr as the Cowardly Lion, plus Norma Desmond, played by Gloria Swanson in Sunset Boulevard (front center).

as $10,000. Once a struggling West Coast artist who studied at the California Institute of the Arts, he won the heart and core of the Big Apple when he moved to Manhattan eleven years ago.

Drawing from the fertile era of the 1920's and 30's, his bizarre brood consists of nostalgic actresses, songstresses, creatures and cultural heroes and heroines. He is confident that his prolificness will never end, since he hasn't enough time to create all his fantasies.

He is noted for his satirical and theatrical subject matter, his attention to detail and his ingenious ability to freeze an animated expression in time. His exhibitions include: Tiffany & Co., 5th Avenue, NYC; The Gallery at Lincoln Center, NYC; Dyansen Gallery, Soho, NYC; River Gallery, Westport, CT.; One Shubert Alley, Century City, CA.; The Eclipse Gallery, NYC. He has done costuming for *Funny Girl*, *Puff the Magic Dragon*, and Wayland Flowers' *Madame*. He also worked with Jim Henson and Associates and at Walt Disney Productions on the West Coast.

215

RON KRON

Sculptor of celebrities, Ron Kron, enjoys a unique position among artists today with seventeen of his original works on permanent display in the Theatre Museum of the City of New York. Not only have Kron's talents been tapped by the museum, but he has also been hired by ad agencies for magazine covers and commissioned directly by such superstars as John Travolta, Angela Lansbury, and Treat Williams to do doll portraits of them as well as Mia Farrow's request for a Woody Allen doll to present as a gift.

Kron with Garbo, Hepburn, Bing, Carmen and more.

Laurel and Hardy (above) are two of the finest male sculptural figures in doll portraiture today. Dietrich (right) is an example of Kron's striking celebrity dolls.

"Backstage" in Kron's studio is where his look-alike stars achieve their true brilliance; Kron is their make-up artist, hair stylist and haberdasher.

Hailed by New York's *Villager* as a "people sculptor extraordinaire," Ron Kron has been sculpting people in miniature since he was a child. An admitted movie buff, his focus mostly pivots around personalities in the entertainment world. In fact, his first sculptural feat, at age ten, was a marionette of his mother's favorite star, Liberace.

The irresistible charm of his dolls comes from a perfect blend of these fundamental ingredients: playful life-likeness, meticulous craftsmanship and exquisite detail—both in the sculptures themselves and in their costuming. Many of Kron's dolls range from ¼ life-size with soft bodies to ½ life-size, jointed figures. Beneath the dramatic overlays of jewels, minks, beaded gowns or denims, his figures are made with latex bodies and cellulose heads, crowned with human hair.

Although his costume designs are usually reproductions of those worn by the actors in movies, Kron also creates his own patterns. Every detail is designed and executed by Kron himself, with one exception—the amazingly realistic eyes made of German handblown glass.

He realized at an early age that dolls evoked a special sense of awe in adults. Remembering back to his high school years, when he and his sister produced puppet shows for young children, Ron said, "We noticed that parents enjoyed our work even more than their kids did." And today, his adult audience continues to be his most enthusiastic fans.

MICHAEL LANGTON

Michael Langton has had two very distinguished honors for his dollmaking, even though he has been sculpting dolls for only five years. His very first doll, Elmer, was commissioned for the film *On Golden Pond*, with Kathryn Hepburn and the late Henry Fonda. The second accolade came from *Esquire* magazine's "1984 Register, The Best of the New Generation: Men and Women Under Forty Who Are Changing America." This award was given to 271 Americans and Langton was chosen along with such luminaries as Steven Spielberg, film director; Steven P. Jobs and Stephen G. Wozniak, co-founders of the Apple Computer; and Sally Ride, the first American woman in space.

Technical proficiency and detailed research have always marked Langton's career whether he was working in commercial art, furniture making, jewelry, sculpture or bronze casting. Although a late-comer to the doll world, he is a member of the National Institute of American Doll Artists (NIADA). His dolls are very popular among doll collectors these days.

Langton's weathered-looking wooden men possess expressions belonging to characters who have found that living in reality can be a tough job. The wood from which the heads are carved also contributes to the aged look of these sculptures. Structurally, they are perhaps the most innovative of the current dolls being made. They are composed of fifty-six separate parts, are about 24" tall, take 200 hours to create and are priced between $2,500 and $5,000.

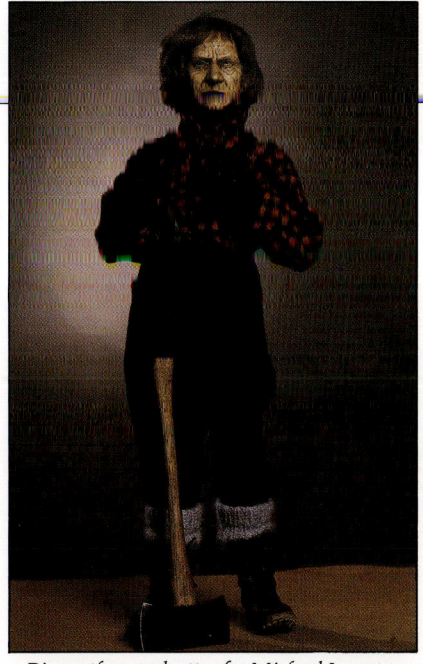

Pierre, the woodcutter by Michael Langton.

The enormous amount of flexibility in Langton's doll construction implores the viewer to participate in maneuvering the dolls into the never-ending range of poses that are possible. Besides the clothing, which is made by a professional tailor, and the fine wigs he purchases, Langton outfits each doll with handmade props like boots, belts, buckles, buttons, glasses and whatever a particular personality might require to finalize its statement. Each doll is totally carved by hand out of fine woods and is a one-of-a-kind sculpture. Currently, Michael is working on a way to reproduce his works through mold-making techniques so that the price will be more affordable and many more collectors will be able to own them.

Gene and Homer wear the maps of reality ingrained on their wooden faces.

Art Dolls

REGINALD FIGTREE

"Old movies are my greatest source of inspiration, recalling a refinement of style seldom seen today," says Reginald Figtree. The San Francisco-born sculptor began his doll career as a puppet-maker, doing large-scale, European-style puppet shows at the San Francisco Museum of Modern Art in the early 1960's.

Since the age of three, Figtree has been involved in dance and theatre work. His solo performance puppet shows were based on Renaissance ballads, Victorian melodramas and the Japanese Bunraku puppet theatre. One puppet show, in which he performed, featured the great "torch" songs of the 1930's and 40's. His dolls are extensions of his puppets, although they are somewhat smaller.

Reginald Figtree's Silk Merchant. Head and hands are sculpted in plastic and baked. The body is a wire armature wrapped in foam, stands 17" tall.

Figtree's love for the theatre and performance does not limit him to the stage work he does. He designs costumes for the theatre as well as women's contemporary fashions.

Each of his dolls is an original Figtree Design, Limited doll. The face and hands are individually sculpted of a bisque-like plastic. Each costume is a unique creation done completely by hand. He does not use a sewing machine. He uses the finest imported fabrics and often features twelve karat gold, pearls and other semi-precious jewels. Figtree is an artisan whose works can be found at doll and crafts shows.

Most of his dolls are long-limbed, fashionable femmes fatales—but we found his representation of Fu Man Chu a fine example of the ever rare male doll.

PAMELA WEIR-QUITON

Fashion conscious Pamela Weir-Quiton began her dollmaking career serendipitously. To placate her parents she attended the school closest to home rather than the school of fashion she really preferred. Her first assignment in her crafts class was to make a toy or doll by laminating woods of different colors together. The rest is history.

Her first doll was 15" high, with geometric shapes, round head and eyes as the only feature. Continuing on with the small wooden dolls, she began dressing them in wood in the trendsetting fashions of designers, i.e. Rudy Gernreich and Betsy Johnson. Apparently, her first love—fashion—had successfully permeated her new love—wood sculpture.

When Weir-Quiton's next assignment required her to make a piece of furniture, the doll motif remained the basis for her sculpture "Sloopy." Reminiscent of Salvador Dali's "Anthropomorphic Cabinet" or "Venus With Drawers," "Sloopy" was a 6' hollow wooden doll with drawers and compartments. After completing two more doll chairs with compartments, she entered the California Design X Show and received national publicity and a commission for a bank lobby which financed her first studio.

Reflecting on how wood has become her main medium, she insists wood chose her, rather than the other way around. She muses that she might have been a tree in a previous life, since the affinity is definitely there. The color and smell of exotic hardwoods remind her that wood is a God-like substance. Weir-Quiton's wooden repetoire includes: 6' ebony mannequins (which come apart to dress), 26" to 36" "Yea Dolls!" derived from the Navajo Indian tradition of sand-painting figures called Yeibichai, the "Liberty Doll," and "Mabel and Her Dancing Cats."

Weir-Quiton's 6' Yea Dolls!, called man-icons, will display fashion when they grace the windows of Bergdorf Goodman in New York.

which were part of large works she designed for playgrounds. Recently, Bergdorf-Goodman in New York City commissioned Weir-Quiton to do a series of 6' mannequins to display their 1986 fashions.

Certainly, no one could have predicted that wood could serve this indominatably creative woman as a medium for fashion. Yet, fashion, historically, is unpredictable and Weir-Quiton is certain to be leading the way in the avant-garde worlds of fashion and dollmaking.

Yea Doll! stands 26" high.

COLLEEN CHAFFEE

Colleen Chaffee is one of the many to be inspired by the wonderful example of Erté. Her first doll, "Claudia," was "created so she could wear pieces such as Erté's." "Claudia" is a moveable bronze doll that achieved much acclaim at the National Institute of American Doll Artists (NIADA) Convention Critique in 1985. Her head, hands and feet are made of bronze, and her mobility is due to the sixty wooden and polymer parts which comprise the rest of her body.

Beginning as a wood carver, Chaffee was employed to design the surfaces for a cabinet maker. Soon she found herself sculpting a series of miniature carousel horses. One of the horses appears in the book *Fine Woodworking Design, Book Three.* Sew-

Pamela with "Mabel and Her Dancing Cat," which was part of her playground series.

Art Dolls

"Claudia" can be arranged in a number of striking poses, due to Chaffee's innovative, patented body design. Claudia's expression is mysteriously haunting because she has no eyes in the darkness behind her lids. Perhaps this doll is the first to be cast in bronze. Chaffee feels her doll is gaining acceptance as a major transitional piece in recognizing dolls as works of fine art.

ing clothes for local dollmakers, while continuing to carve in wood, she eventually found the softer more pliable medium of wax a more attractive medium. After a visit to a doll museum in Vermont, she came up with an original idea to make a wax doll and have it cast in bronze.

MKR Designs
MARILYN K. RADZAT/DOLLMAKER
MARTHA D. KENNEY/COSTUME DESIGNER

Marilyn K. Radzat is most philosophical about her involvement with dolls. Magic and joy are the spirited foundation for all her creations. Working as a sculptor for the last eighteen years, she says "My joy comes in seeing what appears through my hands—a friend who has come to visit..."

Her current medium is a ceramic-type material which can be cured to porcelain-hardness. Each piece is hand sculpted, cured, cooled, and painted with a wash of oil paints giving the skin tone a natural glow. Once the head and hands are complete, Martha Kenney, also a highly skilled artist and craftswoman, creates the soft sculpture body

The "Wizard" is a result of the collaborative effort of sculptor/artist Marilyn K. Radzat and costumer Martha Kenney of MKR Designs.

around a wire armature with fiber-fill. Kenney then creates a costume from antique materials and trims, dating back to the turn of the century. The flexible sculpture is agile enough to frolic when arranged with arms up-raised and legs free for dancing.

After Kenney finishes the costume she returns the robed figure to Radzat who adds the final magical ingredients such as natural furs, feathers, a special wand or a ribbon to complete the spell and set the Wizard free.

PHOTO CREDITS

All photos by Bob Dennison except for those listed below.

King Tut, Lee Doltin, 20
Court Dolls, Clifford Bond, 24
Lily Langtry, Stephanie Farago, 30
Lanternier, Sandra Lee Kaplan, 32
Reclining wax, Sandra Lee Kaplan, 32
Le Fitte-Deserat, Sandra Lee Kaplan, 32
Betty Boop, Sandra Lee Kaplan, 39
Success, Sandra Lee Kaplan, 75
Cadet, Sandra Lee Kaplan, 76
Lindbergh, Sandra Lee Kaplan, 76
Teddy Roosevelt, Sandra Lee Kaplan, 77
King and Queen, Sandra Lee Kaplan, 78
Kitschy Couple, Sandra Lee Kaplan, 78
Wax Peddlar, Sandra Kaplan, 79
Rosebud, Sandra Lee Kaplan, 82
Ivory Sisters, Sandra Lee Kaplan, 84
Latin Lady, Sandra Lee Kaplan, 84
Othello, Sandra Lee Kaplan, 87
Agatha Christy's Inspiration, Sandra Lee Kaplan, 94
Devil, Sandra Lee Kaplan, 101
Maxine Miller, Sandra Lee Kaplan, 120
Marlene Dietrich Doll, Sandra Lee Kaplan, 128, 129
Country Girl, Sandra Lee Kaplan, 136
Pleasant Peasant, Sandra Lee Kaplan, 136
Not A Creature Was Stirring, Sandra Lee Kaplan, 137
Huret Jester, Sandra Lee Kaplan, 161
Solemnity, Sandra Lee Kaplan, 163
Classic Clown, Sandra Lee Kaplan, 169
Facade of Musée National, Stephanie Farago, 180
Dalton with ostrich fan, Sandra Lee Kaplan, 182
Hermaphroditic Court Doll, Sandra Lee Kaplan, 183
Artist-Poet, Stephanie Farago, 184
Equilibriste, Stephanie Farago, 184
Tiffany Lamp with Ivory Nude, Sandra Lee Kaplan, 197
Nude Paper Dolls, Sandra Lee Kaplan, 198
China with Fan, Sandra Lee Kaplan, 199
China Eve, Sandra Lee Kaplan, 199
Wood Nymph, Stephanie Farago, 205
John Noble doll, Sandra Lee Kaplan,, 214
Wotan, M. J. Magri, 215
Craig's bizarre brood, M. J. Magri, 215
Portrait of Van Craig and his characters, Christopher Morris, 215
Pierre, and Gene and Homer, Thom Hindell, 217
Man-icons, Pamela Weir-Quiton with Mabel and Tiger, and Yea Doll!, Myron Moskwa, 219
Back cover photos, Myron Moskwa
Bob Dennison

BIBLIOGRAPHY

Allan, Tony. The Glamour Years Paris 1919 — 40. Greenwich, Ct.: Bison Books Corporation, 1977.

Anderton, Johana, Gast. Twentieth Century Dolls from Bisque to Vinyl. North Kansas City, Missouri: The Trojan Press, 1971.

Axe, John. The Encyclopedia of Celebrity Dolls. Cumberland, Maryland: Hobby House Press, 1983.

Baschet, Roger. Le Monde Fantastique De Musee Grevin. Paris: Tallandier Luneau-Ascot, 1982.

Batterberry, Michael and Ariane. Fashion The Mirror of History. New York: Crown Publishers, 1977.

Bettelheim, Bruno. The Uses of Enchantment. New York: Alfred A. Knopf, 1976.

Buchholz, Shirley. "Lenci–Her Story." Doll News, Volume XXIX (Number 3, Summer, 1980): pg. 10-13.

Buchholz, Shirley. "Lenci–Her Story (Part II)." Doll News. Volume XXIX (Number 4, Fall, 1980): pg. 29-33.

Buchholz, Shirley. "Lenci–Her Story (Part III)." Doll News, Volume XXX (Number 1, Winter, 1980): pg. 26-30.

Coleman, Dorothy S. Lenci Dolls. Riverdale, Maryland: Hobby House Press, 1977.

Coleman, Dorothy S., Elizabeth A., Evelyn J. The Collector's Encyclopedia of Dolls. New York: Crown Publishers, 1968.

Coleman, Dorothy S., Elizabeth A., Evelyn J. The Collector's Book of Doll's Clothes. New York: Crown Publishers, 1975.

Day, Lillian. Ninon A Courtesan of Quality. Garden City, New York: Doubleday & Company, 1957.

Doin, Jeanne. La Renaissance De La Poupée Francaise, Gazette Des Beaux-Arts.

Erté. Things I Remember. New York: New York Times Book Company, 1975.

Franklin, Joe. Classics of the Silent Screen. New York: The Citadel Press, 1959.

Gelatt, Roland. Nijinsky The Film. New York: Ballantine Books, 1980.

Goodspeed, Sally. Boudoir Dolls and Their Era 1910 — 1940 A Report, 1982.

Hall, Carolyn. The Twenties in Vogue. New York: Condé Nast Publications, 1983.

Hillier, Mary. Pollack's Dictionary of English Dolls. New York: Crown Publishers, 1982.

Johnson, Robert A. SHE!. King of Prussia, Pa.: Religious Publishing Co., 1976.

King, Constance E. Antique Toys and Dolls. London: Cassell Ltd. 1979.

King, Eileen, The Collectors History of Dolls. London: Robert Hale Ltd., 1977.

Kirstein, Lincoln. Nijinsky Dancing. New York: Alfred A. Knopf, 1975.

Lafitte-Houssat. Troubadours et Cours D'Amour. Paris: Presses Universitaires de France, 1950

Malipiero, Gianfrancesco. Maschere della Commedia dell'Arte. Bologna: Capitol.

Marion, Frieda. China Half-Figures Called Pincushion Dolls. Paducah, Kentucky: Collectors Books, 1974.

Mitford, Nancy. Madame de Pompadour. London: Hamish Hamilton, 1954.

Mondadori, Arnoldo. Egyptian Museum Cairo. Italy: Newsweek Press, Inc., 1969.

Morella, Joe and Epstein, Edward Z. The "It" Girl. New York: Delacorte Press, 1976.

Noble, John. A Treasury of Beautiful Dolls. New York: Weathervane Books, 1978.

Payne, Frances. The Story of Versailles. New York: Moffat, 1919.

Proverbio, L. Lenci Le Ceramiche. Torino: Edizioni Tipostampa, 1978.

Scagnetti, Jack. The Intimate Life of Rudolph Valentino. New York: Johnathan David Publishers, 1975.

Storey, Robert S. Pierrot: A Critical History of a Mask. New Jersey: Princeton University Press, 1978.

Tate Gallery. The Pre-Raphaelites. London: Tate Gallery Publications, 1984.

Terry, Walter. Isadora Duncan Her Life, Her Art, Her Legacy. New York: Dodd, Mead & Company, 1963.

Tinayre, Marcel. Madame de Pompadour. London: G. P. Putnam's Sons, 1926.

Von Boehn, Max. Dolls and Puppets. Great Britain: McKay Company

INDEX

Accessories, 192
Adorée, Renée, 40, 41
American Doll Companies, 48
American Doll Company Anecdotes, 48
Anita dolls, 38, 82, 103
Antoinette, Marie, 22
Art Deco, 5, 32
Art doll, 5, 33
Automata artists, 183
Baker, Josephine, 64, 69, 110, 132
Bakst, Leon, 31
Ballets Russes, 31
Baleze, 25
Bara, Theda, 37
Barbier, 26, 150
Bartel, 30
Bechoft, 82
Beltrami, Nillo, 113
Bernhardt, Sarah, 29
Bertetti, Clelia, 113
Berzoini, Lino, 113
Betty Boop, 39
Botticelli, 18
Boucher, 25
Boudoir, 27
Boudoir diplomacy, 27
Boudoir dolls, 16, 27, 40, 82, 85, 105, 111
Bow, Clara, 37, 38
Bramble, 206 — 207
Brassiere, 31
Bunraku, 212
Byrd, Charlotte, 207, 208
Cabala, 150
"Carton," 23
Caruso, 153
Ceasar, Julius, 19
Chaffee, Colleen, 219
Chaplin Charlie, 17, 62, 63, 153
Chessa, 113
Clear, Emma, 81
Clown, 150
Cocotte, 30
Cocteau, 33
Columbine, 5, 27, 152, 166, 190
Commedia dell'Art, 28
Corset, 31, 39
Court dolls, 24, 183
Craig, Van, 214
Crash of 1929, 48
Crawford, Joan, 17, 37, 41
Creche, 84
Cro-Magnon, 5
Cubeb Smoker, 5, 48, 50, 51
Cyntherea, 39, 41
Da Milano, Guillio, 113
De Abate, Teonesto, 113
Deco dolls, 46
Deco/European Doll Makers, 45
Del Rio, Dolores, 41
Depression, 16
Descartes, 25
Deschamps, 183
Deval, Brigette Starzcweski, 206

Diaghilev, 31
Dietrich, Marlene, 17, 40, 41, 55, 83, 111, 128, 129, 216
Doll accessories, 46, 47
Dollmaking (history of in Europe), 28
Dolls and Puppets, 17
Dressel-Kister, 28
Dubreau, 173
Dudovich, Marcello, 113
Duncan, Isadora, 31
Dürer, Albrecht, 20
Erté, 7, 8, 9, 32, 210, 211
European Doll Anecdotes, 45
European Doll Companies, 45
Fabergé Eggs, 23
Fadette, 50, 89, 203
Fertility, 17 — 18
Figtree, Reginald, 218
Flapper era, 33
Fool, The, 150, 153
Formica, Claudia, 113
Franklin, Benjamin, 25
French Revolution, 22
Freud, 33
Fuller, Loie, 31
Funerary figures, 19 — 20
Gaichet, Marcel, 208
Garbo, Greta, 17, 216
Gish sisters, 41
Glynn, Eleanore, 38
Goddess, 18
Goebels, 28
Goldsmith, Sara, 214
Grande, 113
Granier, 30
Greek mythology, 18 — 19, 21
Grödnertal dolls, 23
Guilbert, Yvette, 21, 29
Hale, Marsha Bentley, 20
Hampelmann, 206
Harlequin, 27, 150, 152, 157, 171, 190
Harlow, Jean, 83
Haskell, Samuel, 48, 51
Henard, Emile, 188
Henri IV, 21, 152
Hermaphroditic court dolls, 24, 183
Histoire de Jouets, 21
Hobbyists, 16
Hori, Horishi, 212 — 213
Idol, 18
Imans, Pierre, 32, 33, 210
Industrial Revolution, 16, 32
"IT," 37, 38
Kane, Helen, 39
Keeler, Ruby, 50
Kenney, Martha D., 220
Kestner, 28
King Edward III, 20
King Edward VII, 30
King Richard II, 20
La Goulue, 29
Ladies Home Journal, 40
Lafitte-Deserat dolls, 31, 32, 40
Lagrenée, 18

Lambert, 183
Langton, Michael, 217
Langtry, Lily, 30
Lanvin, Mme., 40
Lauder, Barry, 39
Le Infant d' Paridis, 153
"Le Minaret," 32
Lenci Company, 25, 110 — 113
Lenci dolls, 40
Lenci catalogue, 110, 111
Lenclos, Ninon de, 24, 29
"Les Poupées Russes," 9
Lifraud, Mlle., 29
Louis XIV, 23 — 24
Louis XV, 27
Louis-Berault, Jean, 153
Lubitsch, Ernst, 41
Madame de Pompadour, 25 — 27, 28
Madonna, 17
Mannequin, 20, 21, 22, 32, 40, 188
Marie de Medici, 21
Marlef, Claude, 29
Marquise de Rambouillet, 24, 29
Mata Hari, 32
Maury Album, 22
Medieval period, 180
Mercure Gallant, 20
Michelangelo, 20
MKR Designs, 220
Moliere, 25
Moore, Colleen, 41
Mother-role, 17 — 18
Mucha, Alphonse, 29
Musée Grévin, 188, 189, 190, 191
Musée National, Collection de Galéa, 15, 21, 179, 180, 184, 185, 186, 187
Museum of the City of New York, 120
Mussolini, 110
Myers, Carmel, 41
Napolean's son, 22
Negri, Pola, 37, 74
Nietzsche, 30, 34, 47
Nijinsky, 31, 153
Ningyo jyourui, 212
Orientalia, 25
Our Dancing Daughters, 37, 41
Othello, 87
Pagliacci, 153
Pan, 150
Pandora, 20 — 21
Paquin, 40
Peck, Lucy, 30
Petrouchka, 153
"Piajare," 24
Pickford, Mary, 37
Pierrot, 27, 115, 149, 150, 151, 152, 153, 159, 163, 166, 167, 175, 190
Pincushion dolls, 28, 46, 199, 201
Playthings Magazine, 40
Poiret, Paul, 31, 40
Polichinelle, 150, 152, 191
Porcheddu, Beppe, 113
Pre-Raphaelite, 30
Primitive dolls, 17

Prince Ranier and Princess Grace, 80
Provost, Marie, 41
Puchinello, 150, 152
Puck, 150, 152
Punch, 150, 152
Pygmalion and Galatea, 18 — 19
Quaglino, Massimo, 113
Queen Anne of Brittany, 21
Queen Isabella, 20, 21
Queen Victoria, 23
Radzat, Marilyn K., 220
Rejane, Mme., 30
Renaissance, 27, 150
Riva, Giovanni, 113
Roaring Twenties, 33, 89
Roccoco, 25
Roosevelt, Teddy, 77
Royal Doulton, 28
Royal Dux, 28
Rubber girdle, 31
Salon, 24
Salon dolls, 16
Scavini, Elena, 108, 110 — 113, 133
Scavini, Enrico, 110 — 111
Schader's China Doll, 211
Schéhérazade, 31
Segond-Weber, 30
Sorel, 30
Steiff dolls, 112
Sturani, Mario, 113
Swanson, Gloria, 41
Tarot, 150
Taverner, Jules, 16
Temple, Shirley, 41
The Blue Angel, 41
Thompson, Martha, 80
Tosalli, Felice, 113
Toulouse-Lautrec, Henri de, 29
Toys and Novelties Magazine, 46
Tutankamun, 19
Tuturlu (Monmartre doll shop), 208
Union des Arts, 30
United Federation of Doll Clubs, 16
Vacchetti, Emilio, 113
Vacchetti, Sandro, 113
Valentino, Rudolph, 5, 41, 56, 57, 127, 128
Veblen, 31
Velez, Lupe, 41
Venus of Willendorf, 17
Versailles, 23
Vichy, 183
Voltaire, 25
Von Boehn, Max, 17
Voo-doo dolls, 18
Watteau, 25
Washington, George and Martha, 81
Weir-Quiton, Pamela, 218
Wilde, Oscar, 30
Willendorf Venus, 17
Windsor, Claire, 40, 41, 47
Withers, Jane, 41
Woodbury Strong Museum, 32, 76, 78, 84, 199
World War I, 33, 46